Biofilms & Dead Zones:
The Microbe-Environment Connection

How Unseen Life Influences the World Around Us

By

David M. Carlberg, Ph.D.

ISBN: 1-4107-4992-4 (e-book)
ISBN: 1-4107-4991-6 (Paperback)

Library of Congress Control Number: 2003093243

This book is printed on acid free paper.

Printed in the United States of America
Bloomington, IN

1stBooks - rev. 11/24/03

Acknowledgments

A huge debt of gratitude goes out to Dr. Peter Green, emeritus professor of biology, Golden West College, Barbara Pogosian, professor of biology, Golden West College, and Barbara Collins, Principal Analyst, Orange County Sanitation District, who with flea-comb precision uncovered numerous lapses in fact, straightened out tangled syntax and eliminated other inexcusable errors within the manuscript. The author takes full responsibility for any missteps that may remain.

Cover Photo: Acid mine drainage from Iron Mountain Mine, California. Soil microbes in mine tailings have converted the iron in pyrite (iron sulfide) into insoluble iron hydroxide, which clouds creek below the mine with an unsightly rust-orange sediment. Blue splotches are deposits of copper salts, also from the mine. Photo: NOAA Restoration Center – Damage Assessment and Restoration Program

Dedication

To Margaret

Table of Contents

List of Figures

List of Tables

Preface

What we microbiologists take for granted is usually a great surprise to the average person. Many aspects of our daily lives involve all but invisible microorganisms like bacteria and fungi. There is hardly a thing we do, eat, drink, wear, or ride or swim in that is not connected with microorganisms in some manner. Most of those connections are fairly harmless, some are actually beneficial, but a few can be harmful in various ways. This range of possibilities also is true when it comes to the microbial connections with many of our environmental problems. The attack by humans on our environment may have begun with the first discarded bit of unwanted food or clothing. But thanks to the nearly unlimited decomposing abilities of the microorganisms that are found throughout our natural world, that refuse was soon converted into nourishment for nearby plants. Eventually the refuse was transformed once again into something edible or functional. Unfortunately the very same microbes that recycled the discarded matter may also be the ones that rotted the clothing off the backs of the humans or spoiled their food before it could be consumed. And if the humans were unfortunate enough to encounter one of the few microbes that are capable of causing infectious disease, the outcome of that connection might have been deadly.

Over the last two centuries, as the industrial revolution gained momentum, human populations increased and living standards improved, natural recycling processes have not been able to keep up with the immense bulk of refuse, both solid and liquid, that has been heaped and drained onto the landscape. The environment has begun to suffer. Add the more recent invention of so-called recalcitrant and sometimes toxic refuse, materials like polystyrene, DDT and PCBs that won't break down readily by natural microbial action. The persistence of these substances in the environment is measured in centuries. Finally, we have the release of billions of gallons of sewage and urban runoff into our lakes, rivers and coastal waters. While much of this waste can be treated with a variety of physical and biological processes to lessen its impact on the environment, the aging infrastructures that must handle the sheer volume of that waste are beginning to break down.

This book was written for those who are concerned with the health of the environment but who might be looking for a better understanding of the many natural processes where microorganisms play a role both as friends and as foes. The book's chapters are organized according to the four elements of the Greek philosopher Empedocles (495-435 B.C.), fire, water, air and earth, as a simple means of separating the four major realms to be discussed.

Besides the present-day impact microorganisms have on our environment, we will also touch upon how microorganisms over the three and a half billion years of their existence have influenced the very chemical makeup of Earth's atmosphere and geology.

The book is not intended to be a comprehensive treatise on microbiology or environmental science. Some basics of the science of microbiology will be presented in the first two chapters, including the impact of microorganisms on human health, followed by three chapters describing some of the beneficial and harmful impacts microorganisms have on the environment. The book is written primarily for non-scientists, but with the assumption that the reader will remember a few simple concepts learned in high school biology.

We've tried to avoid listing technical and scientific references, since the average reader may not have easy access to them. Instead we have included citations to popular books and periodicals such as Scientific American and National Geographic that are available in most public libraries. Relevant web sites are also included, many of which do contain scientific references. Unfortunately web sites tend to be volatile. The author apologizes if a particular web site comes up "unknown" or otherwise inaccessible. Popular search engines such as Yahoo and Google are also enormously useful in finding web sites concerning environmental microbiology, but be forewarned. Entering the single search term "microbiology" on Google will yield over 1.3 million hits.

We have made one exception to the no technical references rule. That is the May 10, 2002 issue of the scientific journal SCIENCE, which contains several enormously pertinent articles on the significance of microorganisms in the shaping of our world.

SCIENCE can be found in most large public libraries and every university library.

Chapter 1

Introduction to the Little Things in Life

From the bays of the American eastern seaboard and the streams and rivers of Appalachia to the beaches of Southern California and the coastal estuaries of Southeast Asia, innumerable examples of environmental deterioration due to the activities of microscopic microorganisms have been described in numerous technical reports and in articles in the popular press. But microorganisms also offer answers to many of our environmental problems. The unique ability of microorganisms to break down pollutants in our soil and water provides us with the means to reverse much of the damage our modern lifestyle has inflicted upon our planet. More and more corporations and governmental agencies are turning to microbes to solve these environmental issues.

What is Microbiology?

Microbiology is the study of microorganisms, bacteria, fungi, algae and other forms of life that are mostly too small to be visible to the naked eye. A few species of microorganisms, or microbes, are larger than 100 micrometers, which is considered about the lower limit of visibility for the average human eye, but the vast majority of microorganisms are much smaller than that. (One micrometer is one millionth of a meter, or 1/25,400 of an inch.) Many microorganisms, such as bacteria, are approximately one micrometer in size, and most viruses measure less than 1/3 of a micrometer.

For many years biologists assumed insects were the most numerous organisms on Earth, but recent studies show that microorganisms are the most numerous, exceeding all other species in numbers as well as distribution. In fact, it has been estimated that microorganisms make up over half of the biomass, that is, the collective bulk of all living things on Earth.

Microorganisms also are the most widely distributed of Earth's organisms. They are found in every habitat where other plants and animals are present. In addition, microorganisms frequently can be found in environments where the conditions are too extreme for any other form of life to survive. Microorganisms are found in many niches where essentially they are the only inhabitants: the steaming hot springs of Yellowstone National Park, at the bottoms of mile deep drill holes, in Antarctic lakes under 50 feet of ice, and in many of the world's arid desert soils. Many bacteria are not quite as exotic. For example, closer to home, a gram (1/5 teaspoon) of common garden soil or a drop of your saliva may contain over 50,000,000 microbes. A square centimeter of your skin may be home to 1,000,000 microorganisms and half the dry weight of human fecal matter is composed of microbes.

Scientific notation

Scientists frequently must deal with very large numbers, such as the distance between stars or the numbers of microbes in a lump of soil. A system using exponents of ten is known as **scientific notation** and it has been universally adopted to express very large numbers. Thus a gram of garden soil may contain 100,000,000 microbes, which in scientific notation is 10^8 ("ten to the eight"). Notice that the exponent (8) shows how many zeros follow the 1 in 100,000,000. If we wanted to express 3,000,000 in scientific notation, it would come out as 3×10^6.

Scientists often compare large numbers by looking at how many times ten, or **magnitudes**, they differ. If one object weighs 10 grams and another weighs 100 grams, one object is said to be one order of magnitude heavier than the other. If one fossil is 10,000 (10^4) years old and another is 10,000,000 (107) years old, they differ in age by three orders of magnitude.

With all the headline news about bioterrorism and West Nile virus outbreaks, microorganisms have been getting a bad reputation. The fact is, the vast majority of the microorganisms that inhabit our bodies and our environment are harmless, and many are actually critical to the survival of life on Earth. Relatively few microorganisms create problems such as causing infectious diseases, spoiling our food or rotting away the wooden foundations of our houses and bridges.

The Good Guys

Many of the millions of microbes that call our bodies home are actually beneficial. Through competition they protect us from invasion by disease-causing microbes, and some that live in our intestines even provide us with vitamins and other nutrients. The benefits of living in a microbe-laden environment and having microorganisms inhabiting our bodies have been demonstrated by observing so-called germ-free animals. These are mice or guinea pigs that have been raised on normal but sterilized food in microbe-free enclosures. The animals show signs of malnutrition, have underdeveloped immune systems and are highly susceptible to infections. Once they are placed in a normal environment and fed normal food, they are soon colonized by microorganisms and their health returns.

Microorganisms are essential to the survival of life on Earth in other ways. They play a central role in maintaining soil fertility through the recycling of minerals. Without microorganisms, plants would soon run out of nutrients and eventually perish. Without plants, there would be no animals. Plants, as you are aware, and certain microorganisms are the only organisms that can carry out photosynthesis. Photosynthesis is using solar energy to convert the simple gas carbon dioxide, water and other minerals into the organic matter that provides all animals the building materials and energy for making body structure. And as a by-product of photosynthesis, life-giving oxygen is released. In water environments, microorganisms provide the crucial link in food chains which connects the mineral world with the living world. Fish and other aquatic species depend on that link for providing them with essential nutrients. There would be no seafood without microorganisms.

On a practical level, several types of microorganism are immensely useful in helping us manufacture thousands of products such as vitamins, foods (pickles, soy sauce, coffee, cheese, wine, beer, vinegar, etc.), food additives (thickening agents, MSG, aspartame, citric acid), antibiotics (penicillin, streptomycin), and many other products that are used daily by billions of people. Finally, in this age of biotechnology, microorganisms have been invaluable aids in our understanding of how genes work and how we can manipulate genes to treat hereditary diseases or to improve food crops. Microorganisms can be genetically customized to help us manufacture significant quantities of therapeutic products such as vaccines, insulin and human growth hormone. Finally, other specially bred microorganisms can help us solve many environmental problems. Numerous examples of this will be described throughout this book.

Food Chains and Food Webs

A **food chain** is a succession of organisms that are involved in the passage of energy and matter from non-living organic material up through to higher animals. One example of a food chain (Figure 1.1) begins with plant debris in a lake that is consumed by the first and most critical link in the chain, microorganisms. These microbes in turn are devoured by larger microorganisms, then small invertebrate animals. The chain continues up through small fish, a fish-eating bird and finally a hawk, each level being consumed by the next higher level.

Figure 1.1. Example of a food chain.
A dead leaf from a plant is decomposed by microorganisms, which in turn are consumed by larger microorganisms. Nutrients from the leaf are then passed up through successive organisms: an invertebrate, a fish, a fish-eating grebe and finally to a hawk. Organisms are not to scale. Each organism is from 5 to over 100 times larger than the organism it consumed.

But the chain doesn't end with the hawk. The chain actually forms a closed circle, for when the hawk dies, matter from its microbiologically decomposed carcass becomes food for plants and the cycle begins again.

Individual food chains do not function independently. A food chain is often interconnected by organisms in other food chains to form a **food web**.

Cells

All living organisms consist of structural and functional units called cells. Most microorganisms are made up of single cells, while higher forms of life, plants and animals, are multicellular, having many cells. The body of an average human adult, for example, consists of over 10^{12} (Remember, that's 1,000,000,000,000.) cells. There are about 200 different kinds of cells in a human body. Each cell has a specific function to carry out, whether it is to metabolize certain chemicals as in a liver cell, to protect us against infection as in the cells of the immune system, to search out and enter an egg cell as in a sperm cell, to contract as in a muscle cell, to conduct an electrical signal as in a nerve cell, or to carry oxygen and carbon dioxide as in a red blood cell. The single cell of a microorganism must carry out all of the functions that are necessary for its existence, and that is one of the reasons why the study of microorganisms is so fascinating. How can one tiny cell do so much?

A (very) Short History of Microbiology

Classic Microbes

The existence of living organisms too small to be visible to the naked eye was predicted long before they were actually seen. Most ancient references to such organisms had to do with their supposed role in disease. The writings of early Greek and Roman philosophers such as Hippocrates (460-377 BC) and Virgil (70-19 BC) often speculated that some invisible airborne "stain" could be passed from one person to another and cause disease. The idea that minute organisms could make one sick gained strength in the middle of the 16th century with the writings of the Italian physician, scholar and poet Girolamo Fracastoro(1478-1553). In his most important work, *Syphilis sive morbus Gallicus*, ("Syphilis, or the French disease", (with apologies to our French friends)). Fracastoro accurately proposed (in verse) that diseases could be transmitted directly by person to person contact, through contact with inanimate objects, and through the air. He often used the words "seeds" and "germs" in his writings. Most science historians believe he meant living organisms of some type, even though no one had seen microbes until a century later.

The causes of disease continued to attract the attention of a number of scientists up through the 19th century. Their research added support to the idea that microscopic organisms could cause disease, but no one had succeeded in making a convincing connection. That was to await the discoveries of the French chemist, Louis Pasteur, to be described below.

We now know through modern research that a few hundred species of microorganisms do cause diseases in plants and animals, including humans, but the rest of the millions of species are harmless, and some, as noted above, are even beneficial.

Seeing Microbes

To see microorganisms one needs a microscope, an optical or electronic instrument capable of greatly enlarging images of minute objects up to a thousand or more times. The optical microscope was invented around 1590, but microscopes apparently were not used to observe microorganisms until much later. Using simple microscopes of his own design and manufacture, the Dutch shopkeeper Antony van Leeuwenhoek(1632-1723) was the first person to see and describe bacteria and protozoa. Beginning about 1673 Leeuwenhoek, an amateur scientist with an insatiable curiosity, prepared hundreds of letters containing descriptions and drawings of the vast assortment of microbes that he observed. He found microorganisms in rain and river water, in his and his dog's tooth scrapings and saliva and other niches of his personal environment.

Most of van Leeuwenhoek's 200 or so richly illustrated letters were published in numerous issues of the widely distributed journal *Philosophical Transactions of the Royal Society of London*. In addition, he privately published another 150 portraits of microorganisms and other microscopic objects. A contemporary of Leeuwenhoek, British naturalist Robert Hooke, carried out extensive microscopic studies of objects around his home, including insects, plants and minerals. In his best-selling book, *Micrographia*, published in 1665, Hooke described some fungi which were growing on leaves he picked up around his home. He assumed they were some type of plant, but did not pursue his observations of microorganisms further.

Oddly, the discoveries of these pioneers of microbiology did not inspire droves of scientists to an interest in microorganisms and to connect them with the causes of disease. That did not happen until the 19th century and culminated with the work of the French chemist-turned-microbiologist Louis Pasteur (1822-1895). Pasteur is often referred to as the father of modern microbiology. While many of his discoveries were based on the works of earlier scientists, what set him apart was the scientific thoroughness with which he approached practical problems. Pasteur used a type of logic known as the Scientific Method.

The Scientific Method and Scientific Research

The **Scientific Method** is a collection of steps scientists use to answer questions. A typical application of the scientific method may go as follows: A scientist begins by making a simple observation, such as "blood is red." The observation is then turned into a question, "Why is blood red?" The scientist then proposes one or more preliminary explanations, hypotheses or models to account for the color of blood. One hypothesis might be: "Since blood is a complex mixture of many substances, it must contain a component that is red." What to do next? One or more experiments are devised to test which if any of the hypotheses is the correct one. An approach might be to find out a way to separate the hundreds of components that make up blood and see if any of them is red. The scientist soon notes that the color of blood is due to the presence of red-colored cells in the blood, appropriately called red blood cells (RBCs), or erythrocytes. After much work it is discovered that within these cells is a substance called hemoglobin, which is what gives the RBCs their red color.

Based on the new facts, the scientist can then reformulate his hypothesis into a theory, "Blood is red because it contains a substance called hemoglobin, which is red." But the story does not end there. As is frequently the case, scientific discoveries result in more questions. Why is hemoglobin red? Through more experimentation, perhaps by other scientists, it is found that hemoglobin contains something called heme, an iron containing substance. The particular molecular structure of heme causes it to appear red. Why does it appear red? By reviewing earlier work by other scientists it is learned that substances exhibit characteristic colors because they absorb certain wavelengths of light and reflect others. Our eyes collect the wavelengths the substances don't absorb and our brain tells us what colors the substances are. Thus the heme absorbs all wavelengths except red, which makes red blood cells red, which make blood red.

Thus one type of scientific investigation is basically a process of making observations, proposing one or more hypotheses to explain the observations and devising experiments to determine which hypothesis is correct. Once a scientist has analyzed the experimental data and feels his or her hypothesis is supported by the data, the hypothesis will be offered to the scientific community, usually in the form of one or more publications in scientific journals. On the other hand, if the experimental results do not support the hypothesis, it is clearly wrong and must be revised.

After an apparently correct hypothesis is presented to the scientist's peers, it is now open to further testing and confirmation by others. Sometimes, due perhaps to newly developed testing methods or technology, a hypothesis that previously had considerable support may be rejected. At other times, a hypothesis that has survived many years, perhaps even centuries of scrutiny may eventually be upgraded to a law or principle, but even these are still open to question and further testing.

Other scientists may follow a different path, one that involves the search for solutions to real problems. Finding causes and cures for various diseases and developing smaller digital memory devices are two examples of this kind of research, called **Applied Research**. In anticipation of large financial returns, private corporations frequently spend large proportions of their income on applied research, amounting to hundreds of millions of dollars for some major companies. But some companies also engage in **Pure Research**, where their scientists carry out investigations without a specific product in mind nor a guarantee of a return on investments. Trying to find the unifying thread among the four interactions of nature (electromagnetic, weak atomic, strong atomic and gravitation) is an example of pure research. Most pure research is carried out at academic institutions, where the Scientific Method is most often practiced.

Several major discoveries have sprung from activities that started out as pure research, like the discovery of insulin. In 1889, when a team of German scientists were conducting research on digestion, an alert assistant found that the urine of dogs whose pancreases had been removed was loaded with sugar, so much so that it attracted flies. Scientists wondered why and in 1922 their curiosity eventually led to the discovery of the hormone that controls sugar metabolism, insulin, and to the treatment of the disease known as diabetes.

The antibiotic penicillin was discovered unexpectedly by another alert scientist, Scottish bacteriologist Alexander Fleming. In 1929 Fleming, while examining some routine bacteriological cultures, noted that a contaminating mold seemed to be producing a substance that killed the bacteria around it. For most microbiologists, a contaminated culture is scorned and immediately discarded, but Fleming grasped the significance of his chance observation and reported it to the scientific community. The mold that produced the substance had the name *Penicillium*, so he called the substance "penicillin". During World War II, penicillin proved to be one of the most effective weapons against wound infections. Penicillin or derivatives made from it continue to save lives to this day. A fact not often recognized is that the discovery of penicillin also was the catalyst that resulted in the intensive search for other antibiotics, which now number in the thousands. Unfortunately most of these have properties that make them unsuitable for use in therapy, but a few hundred have made it to our pharmacists' shelves for the treatment of infectious diseases. The uncertain future of antibiotic therapy will be examined in Chapter 2.

Thus scientific research is conducted for a variety of reasons. Some kinds of research promise important discoveries and great financial gain while others do nothing more than feed the natural curiosity of scientists.

The Beginnings of Modern Microbiology

In the second half of the 19th century, discoveries involving microorganisms occurred at an accelerated pace. A few examples of the many advances in microbiology made during that period are: the isolation of the bacterium that causes diphtheria, a leading cause of death among young children, and the development of a protective vaccine against it; the discovery of the bacterium responsible for transforming wine into vinegar; the development of our understanding of the role of microorganisms in the conversion of sugar into alcohol and the recognition of the need for disinfectants and antiseptics to prevent infections. Microbiological research flourished in the later decades of the 19th century and culminated in the laboratory of a Frenchman, Louis Pasteur. Pasteur, while trained as a chemist, found himself occupied with a wide variety of microbiological problems. These are just a few of Pasteur's more significant contributions:

•He verified the long-held idea that microorganisms can cause infectious diseases, a principle generally referred to as the "germ theory of disease";

•He extended the work of the 18th century British physician Edward Jenner that showed that microorganisms can be used as vaccines to prevent the diseases they cause;

•He explained that infectious diseases can be transmitted "vertically", that is, from generation to generation via eggs;

•He discovered that microorganisms are involved in both the production of products such as wine and cheese as well as their spoilage; and

•He devised the method now referred to as "pasteurization" in which certain food products can be treated to eliminate spoilage and disease-producing microbes.

Other contemporaries of Pasteur such as the German physician Robert Koch (1843-1910) began similar studies on the role of microorganisms in disease. Koch is best known for establishing what are referred to as **Koch's Postulates**, a set of rules to determine whether a specific microorganism is responsible for a specific infectious disease, the microbiological version of the Smoking Gun. Koch's postulates assumed that a single organism may be responsible for a particular infectious disease, but microbiologists have recently discovered that some infections may be caused by the interaction of two or more microbes. Known as **polymicrobial diseases**, they may involve bacteria, fungi, viruses or various combinations of these organisms. Examples of polymicrobial diseases include periodontal disease, some forms of hepatitis, and certain respiratory infections.

The work of Pasteur, Koch and many others initiated an avalanche of interest in microbiology. By 1880, much of the techniques and terminology that microbiologists use today had been created. By the opening of the 20th century the term microbiology had been coined and microbiology had become a firmly established branch of biology. Universities in Europe and North America began offering courses and awarding degrees, first in bacteriology and then as our knowledge of other microbes expanded, in microbiology. Actually the University of Wisconsin and Harvard University began offering courses in bacteriology as early as 1885. Presently. at least one course in microbiology is offered by nearly every college and university in the United States

The official professional organization of microbiologists in the United States is the American Society for Microbiology (ASM). It is the oldest, and with over 42,000 members worldwide, the largest scientific organization in the world that focuses on a single biological discipline. It publishes a number of scientific journals and sponsors numerous meetings covering over 26 sub-specialties of interest to its members. Governmental agencies and congress often turn to the ASM for advice concerning food safety, public health and bioterrorism.

Types of Microorganisms

Microorganisms are generally organized into five major groups. They are:
bacteria
fungi
algae
protozoa
viruses

Bacteria
The bacteria (singular = bacterium) are the simplest of the microorganisms (except for viruses), but that is not to say they are simple. A single bacterial cell is capable of carrying out a couple of thousand chemical reactions, all of which are assumed to be necessary for sustaining its life. In some cases these chemical reactions are unique to bacteria; no other organisms can perform them. Many bacterial species (but not all) are preeminently metabolically

independent. That means they can synthesize all or nearly all of their complex nutritional requirements (amino acids, vitamins, etc.) from the simple sugar glucose and a few inorganic salts such as ammonium sulfate and sodium phosphate. In contrast, a more complex cell, such as a human liver cell or brain cell, depends on an external supply of dozens of complex nutrients that are provided by the body's blood stream, nutrients which must come from our diet.

Bacteria are capable of exchanging genetic material (DNA) between themselves in at least three ways, by processes known as transformation, conjugation and transduction. In contrast, plants and animals have, with minor variations, only one way in their repertory to pass their genetic information to later generations. In **transformation**, a bacterium's DNA may leak out of its cell and be taken up by another bacterium. During **conjugation**, bacterial cells become physically attached to provide opportunity for DNA to be transferred directly from one cell to the other. (Was sex invented by bacteria? You decide.) **Transduction** involves a virus acting as a UPS delivery truck, carrying DNA from one cell to another. In each of these cases, once the DNA has entered the recipient cell, the DNA may then become part of the cell's own genetic data, possibility giving the cell new capabilities. Contrary to the rules of sexual engagements normally practiced by plants and animals, genetic exchange by bacteria need not occur between members of the same species, or even closely related species. The understanding of these processes has been the foundation of one of the most significant developments in biology, genetic engineering.

If one measures success by two criteria, longevity and sheer numbers, the bacteria are the most successful organisms on Earth. Earth is estimated to be about 4.6 billion years old, and scientists have discovered what appear to be fossil remnants of microbes that lived around Earth's one billionth birthday. Whether life originated on Earth or was transported here via meteorites or spaceships from another planet is unknown, the evidence remains that the 3.6 billion year old life forms appear to be similar to what we now know as bacteria. Since these fossilized organisms seem to be somewhat advanced in development, one must conclude life actually originated much earlier.

In 2002, a group of scientists disputed the claim that the 3.6 billion year old fossils were bacteria but rather were simply chemical artifacts. Regardless of the uncertainty of exactly when life first appeared on Earth, it obviously happened. During the early development of Earth, the physiological activities of the descendants of these early organisms gradually changed the basic composition of our atmosphere from what it was when Earth was first formed to what it is today. This aspect along with the impact microorganisms had on Earth's geochemistry is discussed further in Chapter 5. All the life forms of our present planet, every plant and animal, appear to have evolved from these organisms.

As mentioned earlier, a typical human adult body consists of over 10^{12} cells. In and on our bodies are about 10^{13} bacteria, mostly occupying our skin, mouth and intestines. Much the same can be said of every animal on earth. In addition, every drop of water in every lake, river and ocean contains thousands to millions of bacteria, as does every spoonful of soil. Add up the number of drops of water in every lake, river and ocean and the number of spoonsful of soil there is on Earth and multiply that sum times (as a start) one million and add the human and animal population of Earth times 10^{13} and you will begin to get a rough estimate of how many bacteria there are on our planet; at least 10^{30}, one microbiologist has estimated (that's one thousand billion billion billion).

The structure of a typical bacterial cell is shown in Figure 1.2a. By comparing that structure with the cell in Figure 1.2b, it is clear that the two types of cells are quite different. The cell depicted in Figure 1.2b is a typical animal cell. An important feature of that cell has to do with its genetic material. The genetic material (DNA) in the animal cell is carried on numerous complex structures called **chromosomes**, which are held in a sac-like structure called the **nucleus.** The nucleus is surrounded by a thin membrane (the nuclear membrane) that separates the nucleus from the cytoplasm.

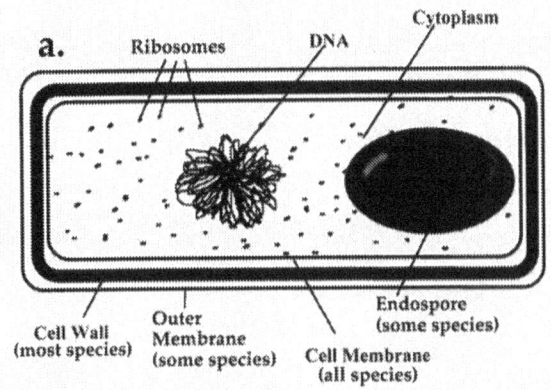

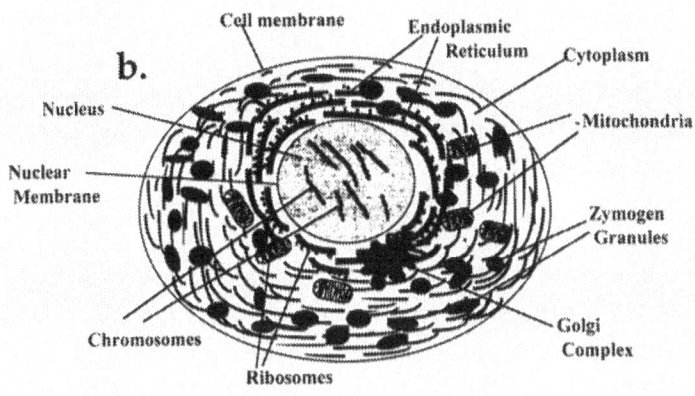

Figure 1.2. a. A typical bacterial cell.
A Prokaryote cell type. b. A typical animal cell. A Eucaryote cell type.

There are other membranous structures, often referred to as **organelles,** throughout the animal cell. Except for one or two membranes that are part of their cell covering, most bacteria have no membranous structures. There is no nuclear membrane; bacterial genetic material consists of a single molecule of DNA that is suspended freely in the cytoplasm. Most bacteria have a heavy, chemically unique cell wall which animal cells do not have. These and some 40 other differences set bacterial cells apart from animal cells and place them in a cell type known as a **Procaryote**. The cell type shown in Figure 1.2b, a typical animal cell, is called a **Eucaryote**. Thus every living cell can be identified with one of the two cell types. The cells of organisms other than bacteria, including all other cellular microorganisms and all plants and animals, are Eucaryotes.

About a dozen genera of bacteria have the ability to form unique internal structures known as **endospores**. These bodies, usually referred to simply as spores, are enormously resistant to harsh environmental conditions such as heat, radiation, chemicals and desiccation (drying). Thus spore-forming bacteria, some of which are responsible for a variety of serious human diseases, are extremely difficult to eliminate and can present a grave health threat (see Chapter 2). Spores are essentially dormant. They show no significant physiological activity and cannot reproduce. The active forms of bacterial cells, that is, those that are capable of reproduction, are called **vegetative** cells.

The procaryotes consist of two separate "domains", **Eubacteria** and **Archaea**(called Archaebacteria for a time). Members of the Archaea look very much like bacteria but have sufficient differences to justify setting them apart from the Eubacteria. Archaea, or ancient ones, are called that because of their proposed close descendancy from Earth's earliest forms of life. This supposition is based in part on the ability of many members of the Archaea to grow in extremely harsh environments such as those possibly present during the early eons of Earth's history. It appears that while the Eubacteria and the Archaea descended from a common ancestor (named the "Last Universal Ancestor" by contemporary microbiologist Arthur Koch), the Archaea are as closely related biochemically to the Eucaryotes as they are to the Eubacteria. The Eubacteria appear to have branched off independently and show only minor connections to Eucaryotes. For the sake of simplicity, further mention of bacteria in this book refers to both Eubacteria and the Archaea.

A discussion of bacteria would not be complete without mentioning those organisms that are able to grow under extreme conditions of temperature, pressure, salinity and acidity. It is not unusual, for example, to find bacteria living in hot springs, caves and deep in the oceans in environments some of which are the equivalent of boiling battery acid. These organisms, most of which belong to the Archaea, are collectively referred as **extremophiles**. Many of the conditions in which these organisms live appear to occur on other planets or their moons, leading scientists to speculate that provided liquid water is present, life on these other bodies may well be within the realm of possibility. Examples of the activities of extremophiles will be brought up in later chapters.

Fungi

Fungi are eucaryotic organisms, some of which are unicellular and others are multicellular (Figure 1.3). Unicellular fungi are commonly called yeasts and the multicellular fungi that grow in connected chains of cell are called molds. To make things interesting, some species of fungi can alternate between looking like yeasts and looking like molds. Of all the microorganisms the molds probably represent the greatest economic impact on agriculture, both positive and negative. Molds are very common in soil and along with the bacteria play an enormously important role in the recycling of nutrients and maintaining soil fertility. That aspect is discussed in Chapter 5. Other molds are major causes of crop diseases, costing billions of dollars to farmers and consumers. Molds are also responsible for mildew and many types of food spoilage.

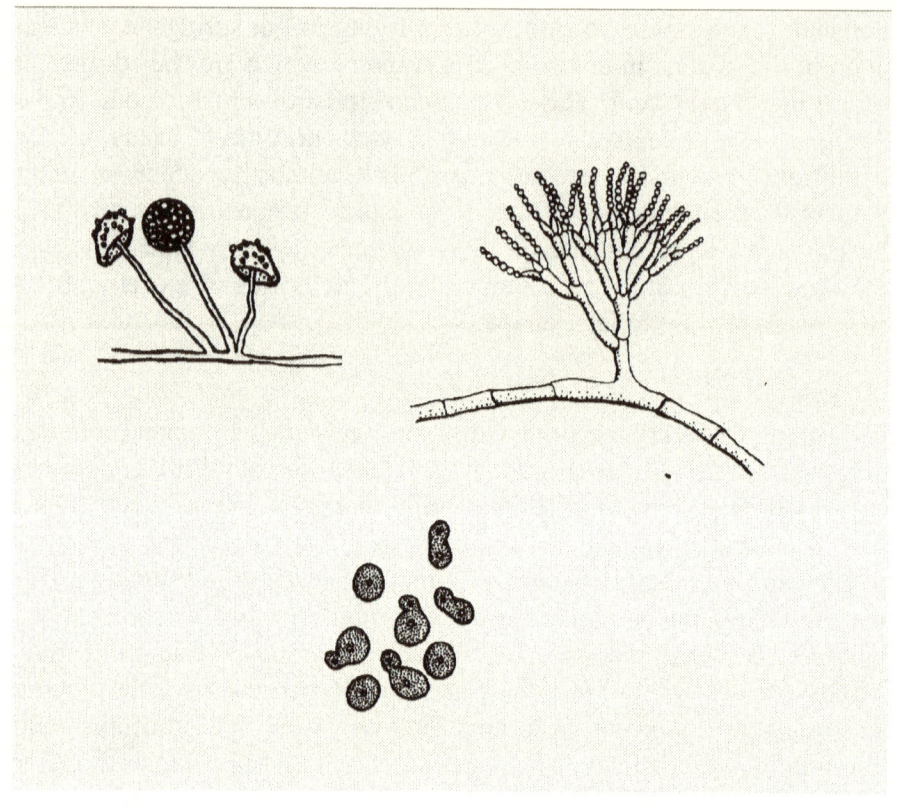

Figure 1.3. Typical molds (above) and yeasts (below).
The yeast cells shown are undergoing reproduction by "budding".

On the positive side, many of our most effective antibiotics such as penicillin are produced by molds, and certain molds are responsible for the production of a number of food products such as cheese and soy sauce. Incidentally, mushrooms, morels, truffles and toadstools are types of fungi. (The Italian word for mushrooms is *fungi*.)

Yeasts are familiar to anyone who has made bread, wine or beer. The ability of yeasts to convert sugar to carbon dioxide and alcohol has been exploited for thousands of years and was first examined in detail by Louis Pasteur in the late 1800s. Genetically-engineered yeasts are the second most common microorganism used in biotechnology (after bacteria) for the production of important pharmaceutical products.

Algae

Algae are usually divided into two major groups, the macroalgae and the microalgae. The macroalgae consist of plant-like organisms often called seaweed or kelp and are found growing in coastal waters world-wide. The microalgae are single-celled microorganisms and are the type one frequently sees growing on the sides of aquariums, swimming pools, tree trunks, soil and rocks along lake edges and stream beds. They exhibit a wide variety of shapes and sizes (Figure 1.4). Algae usually require high levels of moisture, as well as light, since like plants, they are photosynthetic. While most microalgae are harmless, a few can cause serious illness in humans and other animals due to the formation of powerful toxins (poisons) (see Chapter 3).

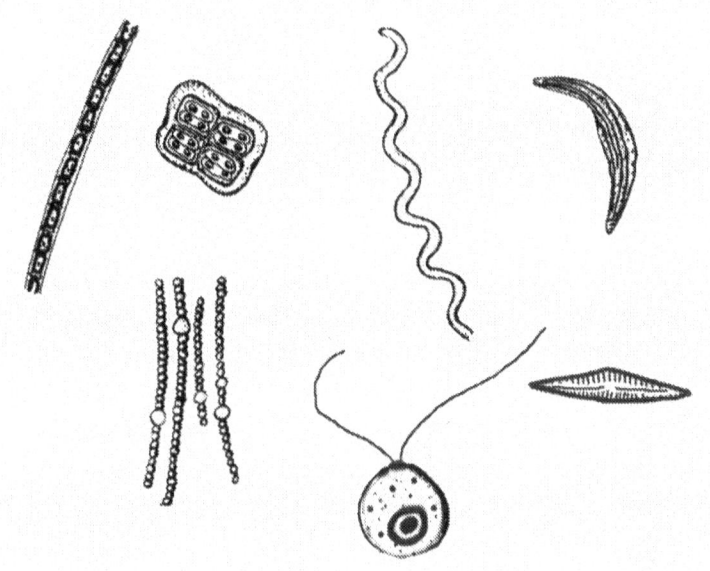

Figure 1.4. Some common microalgae.

Protozoa

The protozoa are the largest of the microorganisms and appear in an enormously diverse range of sizes and forms (Figure 1.5). Many human diseases such as malaria, giardiasis, amebic dysentery and African sleeping sickness are caused by protozoa. In spite of these dire examples, the vast majority of protozoa are harmless and many are important links in food chains.

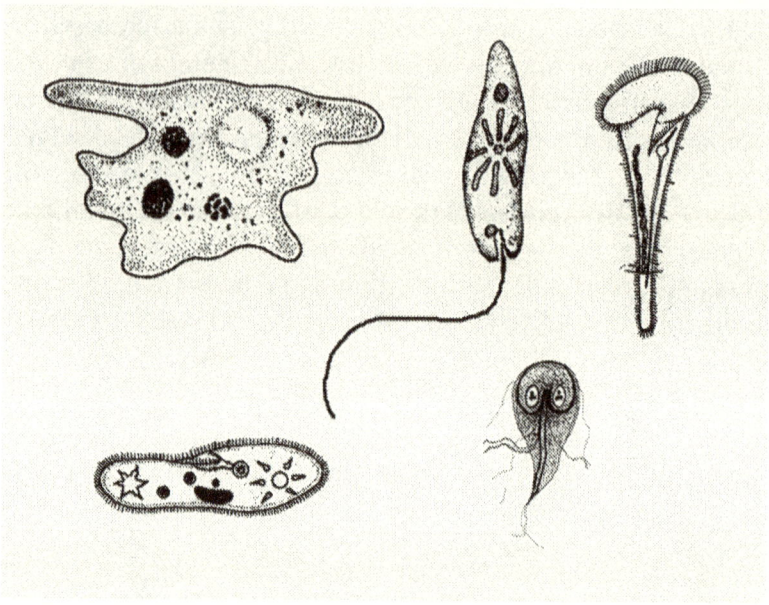

Figure 1.5. Common protozoa found in freshwater.
Top row: amoeba, Euglena, Vorticella. Bottom row: Paramecium, Giardia. Organisms are not drawn to scale.

Like algae, protozoa generally require moisture and are most often found in the intestines of animals, in lakes and rivers or in moist soils where they dine on bacteria, yeasts, algae and other protozoa.

Viruses
Viruses are non-cellular microorganisms (Figure 1.6). They do not have any of the major characteristics of cells. They possess neither cell walls nor membranes nor independent metabolism. The only major feature they have in common with other microorganisms is that they contain genetic material in the form of either DNA or RNA, but a virus never has both DNA and RNA like other microorganisms. Viruses are referred to as obligate intracellular parasites, meaning they must grow within a living cell, called the host. They reproduce by a process known as synthesis and assembly.

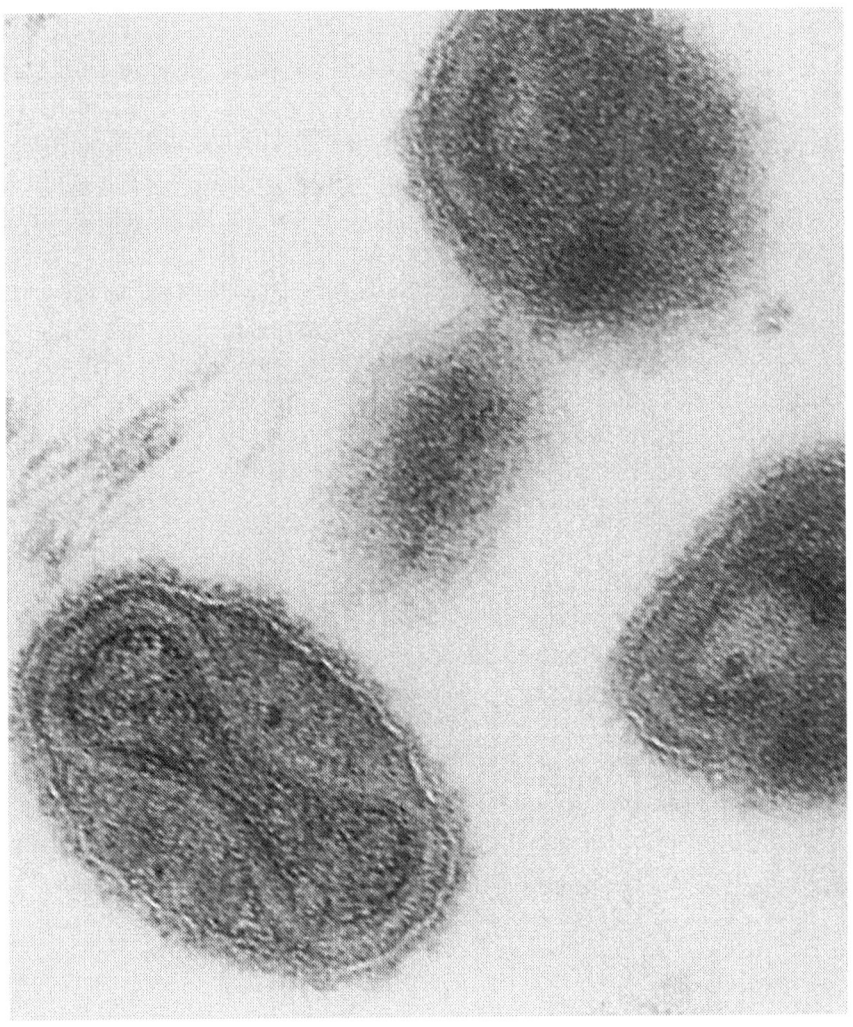

Figure 1.6. An electron micrograph of a smallpox virus.

It measures about 0.24 X 0.27 micrometers, making it one of the largest viruses known. Photo: Centers for Disease Control/Dr. Fred Murphy and Sylvia Whitfield.

When a virus infects a susceptible cell, the genetic material of the virus directs the host cell to stop making host products and begin manufacturing virus parts. In a short time the parts seen to accumulate in the host cell and are then assembled into whole viruses. Soon the host cell may be laden with as many as 10,000 viruses. The viruses are eventually released from the host cell to infect nearby cells, and the cycle begins again. The host cell may or may not be killed by the infection, depending on the host and type of virus, but the functioning of the host cell is almost always altered in some way. In a plant or animal, when enough cells are involved in the virus infection, signs and symptoms of disease may begin to appear. All levels of living organisms are infected by viruses. Plants, insects, animals, fungi, even bacteria are infected by these microorganisms.

Are viruses alive? Biologists have some very specific criteria for defining life. In order for an object to be "alive", as a minimum it must:

1. have a cellular structure
2. grow
3. metabolize
4. exhibit homeostasis-the ability to control its activities to keep everything in balance
5. be capable of some form of movement
6. respond to stimuli in the environment
7. reproduce

Except for criterion number 7, viruses fail on all other counts, at least without the help of a host cell, and they can't pass number 1 under any circumstances. Based on these classical criteria, we can conclude that viruses are not alive.

The origin of viruses has provoked debate among microbiologists since viruses were discovered in 1892. The most common, but not the only, explanation is that viruses "spun off" from the genomes of existing cells and became obligate parasites. Regardless of their origins, viruses surely represent a unique and interesting group of microorganisms.

Viruses range in size from about 0.3 micrometers down to about 0.012 micrometers. Since these dimensions are smaller than the wavelengths of "visible" light that the human eye perceives, individual viruses usually cannot be seen in ordinary laboratory microscopes. To see viruses, one must use an electron microscope, which uses a beam of electrons as the source of illumination. Electrons have wavelengths that are considerably shorter than visible light, making it possible to visualize viruses in the electron microscope. An electron microscope displays images on a screen much like a computer monitor. The images can then be photographed or stored in digital form.

Biological Nomenclature

As early humans became more and more aware of their environment, they began to apply specific names to the plants and animals around them. This had the practical value of making communication more efficient, since it became apparent "large shaggy thing" could refer to a number of objects, including certain members of the family. More recently when scientists began to communicate across international boundaries, confusion often resulted, for references were usually made to plants and animals by their local names. Thus an American biologist might have difficulty explaining his research on Swainson's hawks to an Argentine colleague, who only knew them as *aguilucho langosteros*. The biologists can get help from an international system of nomenclature (naming of things) that was actually established back in the 16th century. In that system, all known plants and animals are tagged with a scientific or binomial ("two name") designation consisting of the **genus** name and the **species** name. Thus Swainson's hawks are now known around the world as *Buteo swainsoni, Buteo* being its genus name, and *swainsoni* its species name. By international agreement, binomial names must be in Latin or "latinized" versions of other languages. And because binomial names are foreign words, they are always italicized or underlined in printed documents.

The system of binomial nomenclature is not simply to help scientists communicate. The system is based on the relatedness of organisms with one another. For example, within the genus *Buteo* are several species of hawk that appear to be related. Some examples are: *jamaicensis* (Red -tailed), *lagopus* (Rough-legged), and *lineatus* (Red-shouldered). Relatedness is based for the most part on how many physical and physiological characteristics organisms share. The science of figuring out how organisms are related to one another is called **taxonomy**.

A system of binomial nomenclature also has been applied to most known microorganisms. *Streptococcus pyogenes* (the cause of "strep throat") and *Escherichia coli* (*E. coli* for short) are two examples of binomial names of familiar bacteria. Occasionally members within a species may differ by one or two characteristics, not enough to justify splitting them into different species, but still requiring recognition. These organisms can be designated as **strains** within the species.

Besides bacteria, other microorganisms such as fungi, algae and protozoa are also labeled with binomial names. Some virologists have tried to develop a system of binomial nomenclature for viruses, but it has not been widely adopted, at least not down to the species level. There are 175 genera of known viruses containing over 4000 members, but in scientific reports viruses still are generally referred to by their common names: mumps, measles, flu, HIV etc.

The Bad Guys: Microbes and Disease

The small fraction of microbes in our surroundings that have the ability to harm us are known collectively as **pathogens**. Pathogens may be found on our skin, in our digestive tract or genitourinary tract at any time as part of the normal microbial community. They also may be encountered in the soil, air, water and food that we are in contact with every day. When pathogens do make us sick they do so usually because they have managed to gain entrance into our body and have overcome the body's defense mechanisms, sometimes at a time when our defenses have weakened due to other conditions such as another illness, malnutrition, drug use or fatigue.

Once the microbes begin to reproduce they release by-products that are either destructive to tissue or outright toxic (known as **toxins**). We can also get sick if we have eaten food in which certain pathogens have been growing previously and released toxins. Regardless of the source, the toxins interfere with the normal functioning of certain organs and eventually a person may show signs and symptoms of disease (see box). The measure of the ability of an organism to cause disease is called **virulence**. Strains of the same organism such as *Staphylococcus aureus* can have differing levels of virulence due to different amounts of toxins or other offensive weapons produced by the organisms.

Disease, Infection, Signs and Symptoms

Signs and symptoms are manifestations of disease. **Disease** is defined as a malfunctioning of an organ or tissue. Disease can be caused by **infection** (growth of a microorganism in the body) or by other causes such as faulty genetic information or exposure to toxic substances. Thus it is possible to have infection without disease, and disease without infection. **Signs** are displays of the presence of disease, such as fever or a rash, that are observable by another person. **Symptoms** are what a person feels that no one else can, like pain or photophobia (sensitivity to light). A **syndrome** is a collection of signs and symptoms.

Natural Resistance and Immunity

Pathogens are usually kept at bay by competition with other microorganisms on our bodies and by various bodily defenses such as natural resistance and immunity, but unfortunately, as we all sadly know, these safeguards occasionally fail us. What is the difference between natural resistance and immunity? **Natural resistance** (sometimes called **natural immunity**) refers to all the inherent, non-specific barriers our bodies possess to ward off invading microorganisms. The unbroken skin and the mucous membrane linings of our oral cavity and intestinal, urinary and genital tracts are often referred to as our first line of defense against infection.

Generally, (there are numerous exceptions) microorganisms must depend on cuts, burns, insect bites and other types of wounds that penetrate these layers before invasion can occur and infection can be established. Also, the body produces a number of protective anti-bacterial chemicals such as **lysozyme**. Lysozyme is present in tears, saliva and nasal fluids and has the ability to dissolve bacterial cells. The respiratory system is lined with cells that have hair-like **cilia** that sweep out small inhaled particles, preventing them from entering the lungs. A major feature of these protective mechanisms of natural resistance is that they are non-specific. They are not directed at any particular organism but are general barriers much like a screen door that keeps out everything from mosquitoes to stray dogs.

Acquired immunity (in contrast to natural immunity) is an enormously complex system that protects the body from invasion by specific organisms or toxic substances. Whenever such material gains entrance into the body, the immune system recognizes it as foreign and counterattacks with a variety of chemicals and cells, the most important of which are proteins known as **antibodies,** and a panoply of specialized blood cells called **lymphocytes**. The purpose of the antibodies is to neutralize the invading organisms or their toxic products and aid the body in getting rid of them. The key point to remember is that the antibodies are directed toward the specific invading organism or substance. For instance, antibodies your body produces against influenza virus will have no effect against tuberculosis bacteria and vice versa.

Macrophages are one type of lymphocyte. They work within both the natural and acquired immune systems. They are on constant patrol throughout the body and have the capability of recognizing trespassers such as viruses or bacteria that they may encounter. The transgressors are immediately ingested by the macrophages in a process known as **phagocytosis**. At the same time the macrophages send out an alarm that triggers other parts of the immune system into action against the specific organism. Once ingested by a macrophage, the microbes are killed with a combination of enzymes and natural disinfectants and then digested. However, some microorganisms, like those that cause Legionnaires' disease, are able to resist the macrophages' inhospitality and can even reproduce while in captivity. Such behavior obviously makes these pathogens much more dangerous.

Another type of lymphocyte, known as **plasma cells**, produces the antibodies that are the major component of the immune system. **Cytotoxic (Tc) cells** and **natural killer (NK) cells** are two examples of cells whose major mission is to eliminate the body's own cells that have been infected by viruses. Tc cells can identify such cells because like a burglar who hangs his coat on the front porch of his victim's house before entering the house, viruses leave signs of their presence on the surface of the cells they have infected. NK cells detect infected cells by other signs. When a Tc or NK cell encounters an infected cell, it doesn't kill the infected cell outright, but releases chemicals that cause the infected cell to commit suicide by a preprogrammed process with the euphonious name of **apoptosis**. Tc and NK cells also are able to search for and destroy cancer cells.

When Immunity Backfires

Occasionally the immune system erroneously identifies cells or tissues of our own body as foreign and launches an attack on them. Known as **autoimmune disorders**, the outcome of such conditions can range from merely inconvenient to fatal. Examples of autoimmune disorders are rheumatoid arthritis, myasthenia gravis, certain types of diabetes, and lupus erythematosus. What triggers this self destructive reaction is generally unknown. Something as benign as a head cold has been known to initiate an autoimmune disorder. The immune system has a self-correcting mechanism that is supposed to prevent autoimmune reactions, but it sometimes fails and the body begins to make antibodies against one of its own parts, often leading to their eventual destruction and possible disease.

Allergy (also known as **hypersensitivity**) is another example of a glitch in the immune system. The immune system inexplicably over-reacts toward normally harmless materials such as plant pollen, cat dander or certain drugs. About 10 percent of visits to doctors' offices is due to allergies. Like in autoimmune disorders, reactions from allergies can range from mild (runny nose, itching) to death.

Vaccines

When you are vaccinated as protection against a specific disease such as polio or tetanus, you are inoculated with a preparation consisting of either whole or parts of microbes or their products. Vaccines against tetanus and diphtheria contain bacterial toxins (poisons) that have been treated to weaken their toxicity. Materials that stimulate antibody production are known as **antigens**. The antigens in vaccines, whether whole microbes or toxins, are made harmless by various methods but they still retain the ability to trigger the formation of antibodies.

It takes several days for the immune system to respond fully to a vaccine, and once the antibodies are made, they will continue to be made as long as the antigen that stimulated their production remains in the body. Antibodies do not last long and once the antigen has been cleared from the body, the antibodies gradually disappear. However, some of the lymphocytes initially involved in producing the antibodies remember how to make them (they are appropriately called **memory cells**). If the antigen (microorganism or toxin) the vaccine was designed to protect you against invades your body at a later time, the body's memory cells immediately recognize the antigen and can respond with antibodies much more quickly and with greater quantity than what occurred the first time the antigen entered the body. You are thus protected by this so-called **anamnestic** or **secondary response** before the invasion can lead to serious disease.

Certain conditions can cause the immune system to fail to respond to antigens. In one instance, the body is unable to make antibodies due to a genetic disease known as **severe combined immunodeficiency (SCID)**. SCID individuals fail to produce the blood cells that make antibodies and must live in a sterile environment. (Remember a 21 year old John Travolta as "The Boy in the Plastic Bubble"?) If SCID patients were exposed to infections that would be mild in people with normal immune systems, the result could be fatal. Modest progress has been achieved with gene therapy in the treatment of SCID, but much more research has to be done. Chemotherapy for cancer and infection with HIV (the AIDS virus) are other conditions that can lead to alterations in the immune system and increased susceptibility to infections.

Inside vs. outside

Why doesn't the body make antibodies against the microorganisms that are growing in our intestine? The simple answer is that inside our intestine is not inside our body. The body can be regarded as a hollow tube, with the mouth being the upper opening, the digestive tract the hollow center, and the anus the lower opening. Thus the body usually does not respond immunologically to organisms in the intestine because that is still outside the body. Organisms usually must penetrate the intestinal wall before the immune system responds. Since many of the microbes in our intestines are beneficial, it is in our best interest not to interfere with their activities.

Emerging Infectious Diseases

The incidence of specific infectious diseases does not remain constant with time. Over the centuries the periodic waves of plague, typhus fever and other infectious diseases that swept through Europe are well known examples of such fluctuations. Many of these disasters resulted in significant political and social changes. In more recent times the development of vaccines and improved sanitation has been a significant factor in controlling once common diseases such as smallpox, polio and typhoid fever, but others seem to pop up and take their places. Infectious diseases that were normally limited to some isolated regions of the world, or were rare or even unknown suddenly show up in our midst. Over 20 such **Emerging Infectious Diseases (EID)** have appeared in the last two decades of the 20th century. Examples are Ebola hemorrhagic fever, West Nile fever and Hantavirus pulmonary syndrome. The virus that is the cause of West Nile fever was first isolated in Uganda in 1937, but since symptoms of West Nile fever until now had been relatively mild, the disease raised little concern. In recent years the virus has suddenly increased in virulence and spread to most states of the United States, infecting over 4000 people and causing nearly 300 deaths. What happened? There are many possible causes for the emergence of these diseases. Here are three:

•changes in climate, making it possible for vectors of the EIDs like mosquitoes to move into areas they normally do not inhabit;

•ease of international travel, allowing an infected individual to transmit a disease half-way around the world in less than a day and long before symptoms appear; and

•increased virulence acquired by the microorganisms due to genetic changes.

Because the occurrence of EIDs is largely unexpected, they are especially dangerous, because physicians usually do not recognize them immediately, resulting in delayed and uncertain treatment.

David M. Carlberg, Ph.D.

Enzymes: How They Work and How They Are Made

A typical microbial cell is capable of carrying out a couple of thousand different chemical reactions. For example, here are some reactions that occur in bacterial cells (some may require additional components or occur in several steps):

glucose –> phenylalanine (an amino acid)

glucose –> acetic acid (the acid in vinegar)

carbon dioxide –> methane (natural gas)

carbon dioxide –> folic acid (a vitamin)

These reactions do not occur spontaneously in cells. The reactions require help in the form of catalysts known as **enzymes** or **biocatalysts**. Every chemical reaction requires an **activation energy** that gets the reaction going. An activation energy is like the heat from the fuse of a firecracker; without it the firecracker remains silent. Enzymes lower the activation energy of chemical reactions so they can start at the moderate temperature of a cell, without the need of a hot fuse. A part of an enzyme known as the **active site** accommodates the starting molecule of the chemical reaction (called the **substrate**) and eases it into becoming the **product**. This occurs with great rapidity. In one minute, a single enzyme molecule is capable of converting as many as a million molecules of substrate into product.

Because every enzyme's active site is usually specific for its substrate, each of the two thousand or so chemical reactions that occur in a living cell normally requires a different enzyme, meaning the cell must manufacture about two thousand enzymes. To save energy and materials, bacterial cells (all cells for that matter) have the ability to turn on and off the manufacture of an enzyme depending on whether it is needed or not. More about that later.

Proteins are the main product of a bacterial cell. Proteins are very long chain-like molecules made up of sub-units known as **amino acids**, of which there are 20 different kinds that are most common. Most of the proteins manufactured by a cell function as enzymes, while others become part of the cell's structure or play other roles in the life of the cell. The ability of a specific enzyme to catalyze a chemical reaction lies in the particular types and sequence of amino acids that make up its structure, much like the letters in this sentence are organized into a sequence that forms words expressing a (hopefully) meaningful thought. Thus each and every enzyme in a cell consists of a specific linear sequence of a couple of hundred or more amino acids. So critical is the exact amino acid arrangement in an enzyme molecule, especially that which makes up its active site, that if any one amino acid (or more) is missing or substituted by another, the ability of the enzyme to function could be destroyed.

The type and sequence of the amino acids that make up the many enzymes of a cell are spelled out in the DNA of the cell. For each enzyme there is a region of the DNA that specifies the exact arrangement of the amino acids in the enzyme chain. That region is usually referred to as a **gene**. DNA consists of long chains of another type of biological building block, nucleotides. There are only four different types of nucleotides in DNA, frequently referred to simply with the letters G, A, C and T. A group of three nucleotides (known as a **codon**) denotes one amino acid. Thus for an enzyme that is 300 amino acids long, there would be a sequence in the DNA that is 900 nucleotides long that codes for that enzyme. But DNA does more than just specify the amino acid compositions of proteins. Additional regions of the DNA may act as control points that determine whether a particular protein is even made at all. Other regions of the DNA may have housekeeping functions such as keeping track of when the DNA should be replicated or when the cell should divide.

The possibility was mentioned earlier that the activity of an enzyme could be destroyed if one or more amino acids in its structure were wrong. How could that happen if the amino acid sequence of an enzyme is encoded in the DNA? This can happen by changing the DNA sequence through an event that is commonly known as a **mutation**. All organisms experience mutations, from viruses to humans. Mutations, though rare, occur randomly and spontaneously and are due to a number of external and internal forces. Various types of radiation and certain chemicals are examples of environmental causes, but the most common source of mutations in cells appears to be errors a cell makes itself when it is replicating or repairing its DNA. DNA is replicated in order to pass on copies of it to daughter cells when the cell divides. If the DNA has suffered a mutation and it is not repaired accurately before the DNA undergoes replication, the mutation is "fixed" and is passed on to future generations.

Not all mutations are harmful. A few may actually be beneficial. Brief mention was made above that microorganisms occasionally acquire increased virulence through a genetic change. Such a change could be caused by a mutation, which might afford the organism the ability to make more toxin, for example, or to resist higher levels of antibiotics. A mutation might alter the organism's susceptibility toward a host's immune system. That is what happens with influenza virus. The virus changes genetically so rapidly from one year to the next that it is necessary to modify vaccines every year to match the current virus strain. Finally, suppose a bacterial cell experiences a mutation, causing an amino acid substitution that improves the activity of a particular enzyme. The mutation gives the cell a slight competitive advantage over other cells and eventually the descendants of that cell outgrow all the other cells. Mutations are one of the forces that drives evolution.

A bacterial cell can be compared to a well-run factory whose main products are proteins. The DNA can be considered as the repository of the master blueprints for the proteins. The site of actual protein assembly is the cytoplasm, on thousands of (to continue the factory analogy) "workbenches" known as **ribosomes**.

Enzyme Switchboards

A bacterial cell has checks and balances, assuring that energy and materials are expended in an efficient and unwasteful manner. For example, some enzymes are not produced until required. That is, their genes can be turned off until a need for the enzyme appears. When a specific enzyme is needed, a "work order" is sent to the gene that codes for the particular protein. The order interacts with a special section of the DNA, a control segment called the **operator,** that is responsible for turning the gene on and off. Once the gene is activated, a copy of it is then made in a process known as **transcription**. The copy is the "working blueprint", known as **messenger RNA** (abbreviated **mRNA**). The mRNA is sent to the cytoplasm (factory floor) where it is placed onto a ribosome (workbench). Enzymes and a myriad of other proteins (assemblers) using smaller molecules (tools) then put amino acids (parts) together in the right sequence to make the protein according to the instructions in the mRNA. This step is aptly named **translation** for it is truly translation in a linguistic sense. The language of the DNA, consisting of 4 letters (the 4 nucleotides), is translated into the language of proteins (the 20 amino acids). In some instances the production of a particular enzyme may not simply be controlled by an on/off switch. Depending on the needs of the cell, the production of an enzyme would be modulated like one controls the flow of water by adjusting a valve.

Millions of copies of the enzyme are turned out by the ribosome in a matter of a minute or so, but the mRNA usually is destroyed after each enzyme molecule is made, so a continuous supply of mRNA must be made available. The combined reactions that go into the synthesis of proteins, the making of the nucleotides for the mRNA, the mRNA, the amino acids, and the proteins themselves, all require an enormous amount of energy. That energy must come from the nutrients the cell consumes, or if the cell is photosynthetic, from light. Because over 90 percent of the energy expended by a bacterial cell goes into making proteins, the cell saves considerable energy as well as materials by limiting synthesis only to those proteins that are needed at a particular time, and in appropriate amounts.

One way cells are able to turn on genes when needed is by sensing the amount of substrate that is available. For instance, some bacteria produce an enzyme called beta galactosidase, which is involved in the first step in the breakdown of the sugar lactose. The lactose concentration in the cell determines if the beta galactosidase gene is turned on or not. In the presence of significant amounts of lactose, the gene is transcribed, mRNA is produced and the enzyme is formed. The lactose inactivates a **repressor**, a protein that normally keeps the gene from undergoing transcription and is thus the equivalent of a "stop order", and the lactose is the "work order" that cancels the stop order. This mechanism where the substrate initiates the formation of an enzyme is known as **induction**. In the absence of lactose, the repressor blocks transcription by attaching itself to the operator much like a large log blocks a railroad track. The overall result is that beta galactosidase is made only if lactose is present.

When sufficient enzyme has been produced, its manufacture may be shut down in several ways. How does a cell know that it has made sufficient amounts of a particular enzyme? One way would be by checking to see how much end product the enzyme has produced. If there is sufficient product present (either what the cell has already made or what is present in the environment), the genes are turned off. If there is little or no product present in the cell, the genes are turned on. In this case, called **end product repression**, product molecules activate a repressor which shuts down the genes. In the absence of product, the repressor is inactive and the genes are free to make mRNA, leading to the formation of the enzymes and ultimately more product. Another way of controlling enzyme synthesis is by blocking the translation step in its tracks. A mechanism known as **attenuation** causes the mRNA to drop off the ribosome when sufficient product is present, preventing synthesis of more enzyme.

In still another system, referred to as **feed back inhibition**, the product of an enzyme may simply inhibit the enzyme directly. This prevents more product from bring made until the concentration of product is reduced to a low level.

All of these various control systems (as well as others not mentioned) may be operating simultaneously in a bacterial cell, and in some cases controlling the same genes. The purpose of all this redundancy is not known. It has been suggested that through these multi-layered control systems bacteria are able to fine-tune the synthesis of certain enzymes in order to keep the metabolism of a cell in perfect balance.

Microbiologists are beginning to discover that when a pathogen senses that it has invaded the body, certain genes associated with infectivity are switched on. That information is exciting news, for it leads the way for the development of a revolutionary new means of controlling infection by targeting the infectivity genes and shutting them down. One approach, already being tried in viral infections, is to inactivate the mRNA specific for the pathogen. With the growing prospect of antibiotics becoming useless in the battle against infection (see Chapter 2), treating infectious diseases at the genetic level may be the next great advance in microbiology.

This short description of the various types of genetic controls for protein synthesis generally applies to bacterial cells, where they were first discovered, but in a broad sense the story is universal: all living cells, including human cells, have various ways of controlling enzyme synthesis and activity. One significant difference between procaryotic and eucaryotic DNA is that the physical structure of eucaryotic DNA is considerably more complex. Ways of controlling which genes are to be transcribed in eucaryotic cells may have to do with how the helical DNA is coiled, for example, or with how certain proteins called histones are attached to the DNA. In the genes of eucaryotic organisms, the specific sequence of nucleotides in DNA that denote a protein is usually not contiguous. That is, like radiator this oyster sentence, chair extra words omelet are inserted football that have purple to be edited out lawn before it makes any pistachio sense. The extra sequences of nucleotides, known as **introns**, are snipped out of the mRNA prior to translation, leaving behind **exons**, the sequences that actually code for the protein. It was this finding that mRNA had to be edited that led to the discovery that not all enzymes are proteins, a bit of biochemical dogma that goes back to the early part of the 20th century. RNA molecules with enzymatic activity are responsible for editing out exons, and they have been named **ribozymes**.

Another notable difference between protein synthesis in the cells of eucaryotes and bacteria has to do with where it actually occurs. In eucaryotes, translation cannot occur until the mRNA is produced, edited and then passed through the nuclear membrane to get to the cytoplasm where it is accessible to ribosomes. Bacterial DNA is suspended freely in the cytoplasm; there is no membranous barrier separating the DNA from the ribosomes. Therefore as mRNA is being produced, ribosomes are able to attach themselves to the mRNA even before the mRNA is completely synthesized. In other words, transcription and translation of a particular mRNA molecule can occur in bacteria simultaneously. Also, mRNA editing usually is not necessary, as introns are rare in bacteria.

The speed with which bacteria can switch a gene on or off for a specific enzyme is one of the secrets of the success of bacteria. Bacteria are able to respond rapidly to changes in the environment, such as the sudden but fleeting appearance of a particular nutrient, an ability that has enormous survival value.

Helpful Enzymes

An immensely important feature of enzymes is that they can operate outside of the cell, a momentous discovery that was made in 1897 by the German chemist Eduard Buchner and for which he received the Nobel Prize in 1907. Enzymes can be extracted from microbial cells and used like any other chemical reagent, provided the proper physiological conditions to support enzyme activity are maintained. There are several good examples of microbial enzymes performing important functions in daily human activities. Certain laundry detergents contain bacterial enzymes that aid in the removal of grease and stains in clothing. Many enzymes help manufacture such diverse products as candy, cheese, beer, the artificial sweetener Aspartame, cotton dresses, antibiotics and other drugs. As we will see in Chapter 5, isolated enzymes also can be used to mitigate environmental pollution.

Microbial Swap Meets

Now that sequencing of the nucleotides in the DNA of bacteria is almost routine, microbiologists have been able to compare DNA sequences of several species. One of the most interesting discoveries coming out of this work is that the sequences of many segments of DNA from unrelated bacteria are either very similar or identical. This suggests that over the 3.5 billion years or so that bacteria have been around, these organisms appear to have been doing a lot of gene swapping. Through a process known as **horizontal gene transfer**, a fragment of DNA from one bacterial cell leaks out of the cell and is taken up by another cell that may or may not be related to the first cell. Rather than break up the DNA into its building blocks to make new DNA, the recipient cell splices the DNA fragment into its own DNA and the fragment becomes part of the cell's permanent genetic information.

The genetic information obtained by the recipient bacterial cell may give the cell additional abilities such as the formation of one or more new enzymes or the acquisition of increased virulence. For example, through horizontal gene transfer normally harmless bacteria such as *E. coli* occasionally have acquired the ability to kill. This fearsome possibility will be described further in Chapter 3.

Horizontal gene transfer also has had an enormous impact on studies of the evolution of bacteria, and indirectly, on our understanding of all life. Primarily it has made tracing origins of bacterial species through DNA sequences extremely difficult. It is like the residents of a large apartment house randomly swapping all forms of identification with one another. Then, if one resident were to be asked to identify herself with a driver's license, she would appear to be Jane Smith, but with a credit card, Alice Jones, and with a social security card, Juana O'Brian. Thus, evolutionary pedigrees constructed for a particular bacterium may vary from one another depending on how the identity of the organism was determined.

Microbial Relations

Parasitism

Parasitism occurs when an organism (the **parasite**) lives in or on a larger organism (the **host**) at the expense of the host. In addition to the pathogenic microorganisms mentioned above, certain plants (e. g., mistletoe) and animals (some types of worms, snails, insects and beetles) also are known parasites. Some parasites, such as human cold viruses, have simple life cycles involving a single species of host and are transmitted directly from host to host. Other parasites like yellow fever virus or the bacterium that causes Lyme Disease, have slightly more complex life cycles that include intermediate hosts such as mosquitoes and ticks, respectively.

Other parasites exhibit enormously elaborate life cycles consisting of numerous and widely diverse intermediate hosts, alternating sexual and asexual reproductive stages, and wildly dissimilar phases. The malaria parasite, *Plasmodium*, fits this description. In the mosquito host the malaria parasite resides in the insect's salivary gland and takes on a form known as a sporozoite. On biting a primate host (including a human) and feeding on the its blood, the insect salivates on the bite wound, releasing sporozoites into the blood stream of the host. The invading parasites first infect the liver and then the red blood cells as the microbes go through various asexual and sexual stages, each of which appears totally different from the others. If the infected host is then bitten by another mosquito (it has to be the right species), and the insect ingests infected blood cells, parasites in the blood cells complete a sexual reproductive cycle in the stomach of the mosquito, then migrate to the salivary gland as sporozoites and the cycle is complete.

Parasites can have narrow **host specificity** or **host range**, such as the human cold virus, or show specificities of varying broadness, parasitizing, for example, only related species, such as members of the cat family. In the case of the fish tapeworm, it infects species as diverse as fish, brine shrimp and humans. Actually most (over 60 percent) of human pathogens also have one or more animal hosts. A bacterium known as *Burkholderia cepacia* is an example of a parasite with an exceptionally broad host range, infecting both humans and plants.

Wildlife biologists are particularly concerned over the fact that many parasites that infect wildlife, including some endangered species, also are found in livestock, common household pets and humans, making the prevention of the spread of these parasites enormously difficult.

The presence of a parasite in a host can have a variety of impacts on the health and fitness of the host. A parasite need not necessarily kill its host outright. In fact, the more successful parasites influence their hosts in more sustained and subtle ways. Since a parasite must derive nutrients from its host, its impact on the host will depend on the host's nutritional state. The health of wildlife in natural habitats is often compromised due to drought, competition for food, injury and other factors. Parasites are known to alter host reproduction and feeding activities as well as induce physical deformities. For example, biologists have recently discovered that the odd deformities long observed in frogs like extra or missing legs may be due to parasites. Parasitism is quite common in the wild and it can have an important impact on the success of plant and animal populations within ecosystems. Add to this the effect of stress on our wildlife due to human activities and we can begin to understand why it does not take large increases in pollution, habitat destruction or alterations of climate to drive certain species to extinction.

Parasitic Origins of Organelles

There is scientific evidence to suggest that at least two membranous organelles found in eucaryotic cells, mitochondria in plant and animal cells and chloroplasts in algae and plant cells, were originally procaryotic cells that either parasitized the ancestral cells of present day eucaryotic organisms, or were ingested by them as food and instead of being digested, survived and began reproducing. Mitochondria are frequently referred to as the power plants of cells while chloroplasts are the solar cells of photosynthetic organisms. Both mitochondria and chloroplasts contain small amounts of DNA, sufficient to code for a dozen or so proteins. These organelles also have their own ribosomes to carry out protein synthesis and are able to grow and divide and at times fuse together to form larger organelles. They act like semi-independent organisms with very limited genetic capability. In other words, they appear to have been originally visitors but became permanent residents.

Mutualism

Parasitism is just one form of interrelationships that exist between organisms. **Mutualism** is a type of symbiotic relationship in which both partners benefit. Examples of mutualism between bacteria and eucaryotes have appeared in a number of habitats. For example, there are certain bacteria that colonize a type of worm known as a nematode. These small (1 to 8 mm) animals are very common in aquatic and soil environments. The bacteria completely cover the outer surface of the worms with a thick layer of cells as well as reside in the worms' intestines. The bacteria are autotrophs, meaning they derive their carbon and energy from inorganic sources: carbon dioxide and hydrogen sulfide, respectively. By analyzing carbon isotope ratios of the bacteria and the nematodes, scientists have discovered that the carbon incorporated in the worms' tissues is essentially the same as the carbon in the bacteria. That means the major source of the carbon in the worms, as carbohydrates, is the bacteria. Remove the bacteria and the worms would starve to death.

Another remarkable example of mutualism between bacteria and animals occurs in certain species of squid. These marine animals have light-emitting organs that provide them with a means of confusing predators. The light emitting organs are sac-like structures on the underside of the squid and are home for luminous bacteria. The light organs of newly hatched squid contain no bacteria, but in a short time the light organs of the juvenile squid become colonized with very specific species of bacteria and no others. The bacteria have the capability of emitting light through a chemical reaction known as **bioluminescence**. The squid's light organ has a lens that focuses the bioluminescent light onto the under-body of the nocturnal-feeding squid like a flood light. Thus the shadowy silhouette of the squid against moonlight that would be seen from below is illuminated by the light organs, causing the animal to blend into the moonlight and become invisible to predators. The intensity of the light appears to be controlled by the squid host by altering the shape of the light organ or through chemical messages from the squid to the bacteria. In return for the protective role played by the bacteria, the squid provides the bacteria with food and shelter.

One of the most complex and bizarre examples of symbiotic relationships between microorganisms and animals occurs in the nests of over 200 species of ants known as attine ants. These insects actually cultivate specific species of fungi for food, an obligate relationship that has been going on for at least 50 million years. The cultivated fungi, however, are frequently under attack by a parasitic species of fungus which can destroy an entire fungus crop of an ant nest and ultimately the ant colony itself. To protect the colony from losing its food supply, the ants also raise a specific species of bacterium that produces an antibiotic that inhibits the invading fungi. The end result appears to be a case of four-way mutualism. The ants depend on their fungal crop as a food source and the bacteria as a defensive weapon. Both the fungi and the bacteria in turn need the ants for food and shelter, and the parasitic fungi depend on the cultivated fungi for their food source.

Growing Microorganisms

In order to study microorganisms they must be propagated or grown in the laboratory. To support the growth of any organism, certain conditions must be met. Some of the more critical conditions that must be provided are the proper nutrients, and appropriate temperature, oxygen level, and pH (acid or alkaline level). The most difficult hurdle in growing a particular species of microorganism in the laboratory is finding the right nutrients. The nutrients offered a given microorganism must match its particular assemblage of enzymes, something that varies widely from species to species and is often unknown. For that reason, it is estimated that only about one percent of the millions of species of microorganism that exist on Earth have been grown in the laboratory. Thus our understanding of microorganisms is based on the study of but a relative handful of species, forcing us to make a lot of assumptions and generalizations. One of the things that makes microbiology so interesting is that microorganisms are constantly surprising us with startling exceptions to these generalizations.

Feeding Microorganisms

A **medium** (plural: media) is a mixture of nutrients that will support the growth of a microorganism. The nutrients may be specifically known, such as glucose or ammonium nitrate, or they may be undefined and supplied as a crude extract of, for example, beef, milk or soybeans. If the medium is a liquid, it is referred to as a **broth**. A vessel containing growing microorganisms is referred to as a **culture**. Sometimes it is advantageous to grow the microorganisms on a semi-solid surface, in which case a solidifying agent is added to the broth. Gelatin and silica gel are sometimes used as solidifying agents, but by far the most common solidifying additive is **agar**. Agar is a polysaccharide extracted from certain seaweeds (macroalgae). At concentrations of one or two percent, agar has the ability to solidify broth to about the consistency of Jell-O. The medium is then simply referred to as "agar". Microorganisms growing on the surface of the agar form visible aggregates known as **colonies** (Figure 1.7). There are a number of advantages from growing microorganisms as colonies. The size, shape and color of the colonies are often helpful in identifying the organism. In addition, populations of microorganisms in food or water samples can be determined by counting the number of colonies that form on the agar when a given volume of sample is spread onto the agar surface. This technique is covered in a later section of this chapter.

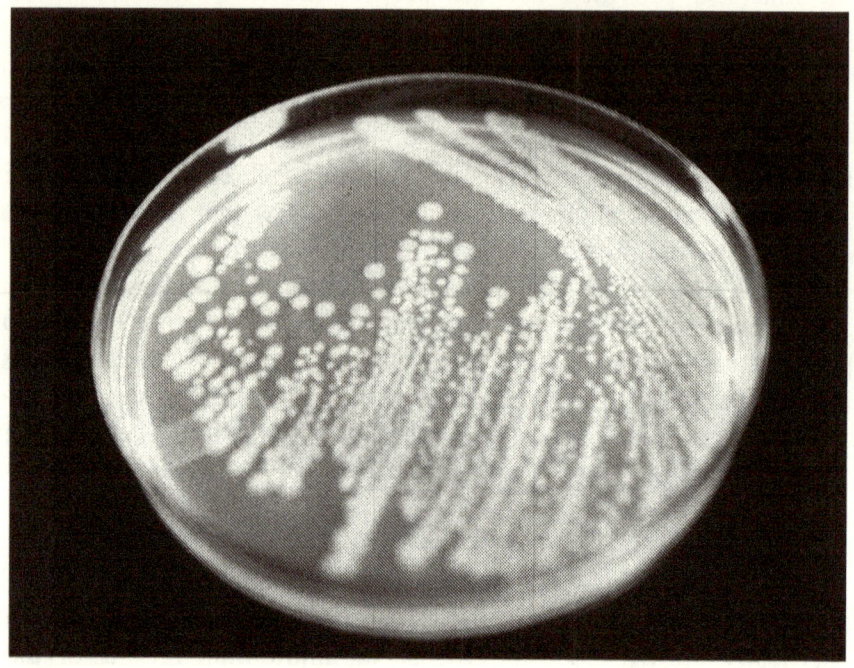

**Figure 1.7. Round objects are bacterial colonies growing
on the surface of an agar medium.**
A colony may contain over 10^8 cells.

By the clever choice of ingredients in a medium, one can produce a so-called **selective medium** in which only certain types of organism are allowed to grow and others are inhibited. This is useful when taking environmental samples to detect bacterial pollution. For example, of the hundreds of species that may be present in a lake water sample, a selective medium will allow only those species associated with pollution (known as **indicators**) to grow. A **differential medium** contains ingredients that allow one to distinguish specific organisms or groups of organisms from others that may be present. For instance, the colonies of bacteria that cause "strep throat" produce very characteristic reactions on agar containing sheep blood, to distinguish them from all the other bacteria commonly found in the throat. Differential media are also used to determine levels of specific indicator bacteria known as coliforms in water samples. The presence of these bacteria suggests fecal contamination. Coliforms produce colonies of a characteristic color on differential media which can be counted and related back to the volume of water applied to the agar. This aspect is covered in some detail in Chapter 3.

Broth is usually contained in culture tubes or flasks. One small tube of a bacterial culture can easily reach a population of 10^{11} cells, some two orders of magnitude greater than the entire human population of the earth. Agar is usually dispensed in Petri dishes to allow a large surface for observing colonies.

Temperature

Temperature is another critical factor in cultivating microorganisms in the laboratory. If one were to plot the growth rates of a variety of bacteria against temperature, one would find that each species prefers a specific **optimum growth temperature**, that is, the temperature at which the growth rate is maximum. (Figure 1.8). Based on their optimum growth temperatures, bacteria can be classified into various groups.

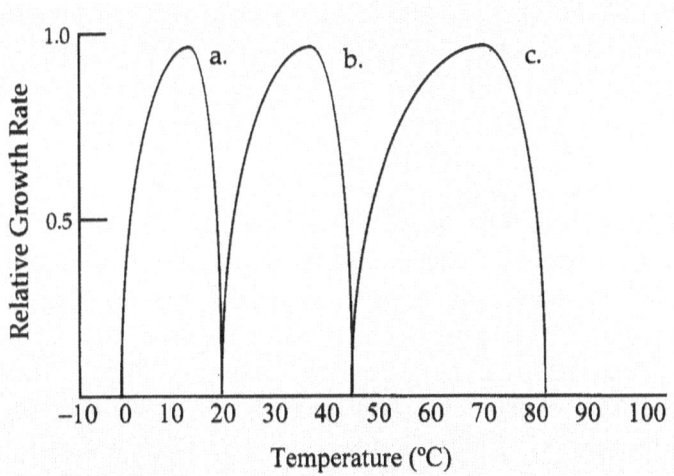

Figure 1.8. Plot of typical growth responses to temperature for three species
of bacteria representing (a) psychrophiles, that grow at 20°C or below, (b) mesophiles, that grow best between 20°C and 45°C, and (c) thermophiles, with maximum growth above 45°C.

David M. Carlberg, Ph.D.

Those bacteria that have optimum growth temperatures between 32°F (0°C) and 68°F (20°C) are called **psychrophiles. Mesophiles** are organisms that grow best between 68°F and 113°F (45°C). These include all human pathogens, which generally have an optimum temperature around 98.6F (37°C), normal human body temperature. Some microorganisms have optimum temperatures above 68°F but can grow at temperatures as low as 32°F, although not well. These are known as **psychrotrophs** and along with the psychrophiles are the organisms that can grow in the cold of your refrigerator, eventually causing food to spoil in spite of the refrigeration. Optimum growth above 113°F is exhibited by so-called **thermophiles**, some of which grow best at temperatures approaching 212°C (100°C, the boiling point of water) or even higher. Thermophiles are often found growing in natural environments such as hot springs, in the soil around active volcanoes, and near hydrothermal vents in the deep ocean. Closer to home they are common inhabitants in spas and hot tubs. In the laboratory, temperature is controlled by cultivating microorganisms in an **incubator,** a type of cabinet with precise temperature control. The determination of growth temperature range as a differential tool is commonly used in testing water for sewage contamination (Chapter 3).

pH
The **pH** of a medium is a measure of its acidity or alkalinity. The pH scale runs from 0 to 14 with 0 being very acid and 14 being very alkaline. pH 7 is considered neutral. The pH of a given medium must be adjusted to match the preference of the organism to be grown. Most common bacteria prefer a pH near neutral, 6.8 to 7.2. Some bacterial species, such as those that are responsible for a highly environmentally damaging condition known as acid mine drainage, do best below 3.0. The bacteria found in the alkaline lakes of Death Valley, California require media with a pH above 8.

During the course of their growth, microorganisms often produce byproducts such as organic acids that cause shifts in pH, eventually inhibiting growth. To offset pH shifts during microbial growth, **buffers** can be added to the medium. Buffers are substances that minimize pH changes in a solution.

Atmosphere

The atmosphere of Earth contains about 21 percent oxygen by volume. While most microorganisms on our planet require gaseous oxygen, there are certain ones that do not require oxygen. For some of these microorganisms, oxygen is actually toxic. Organisms that require the presence of oxygen are called **aerobes** while those that do not require it are known as **anaerobes**. Some microorganisms can grow either in the presence or absence of oxygen. Those are **facultative anaerobes**. Organisms that must grow in the absence of oxygen are called **strict** or **obligate anaerobes**, and these are the ones that are killed by oxygen. Why is that?

The function of oxygen in a cell is to eliminate electrons that are the by-products of respiration. Oxygen is thus referred to as an **electron acceptor**. In respiration, energy is derived from chemicals through a biochemical process known as oxidation, meaning removing electrons. The energy associated with the electrons is used to drive other metabolic reactions in the cell. Like orange skins after juice has been squeezed from the fruit, the electrons that are left after they have given up most of their energy must be discarded. Aerobes, including all animals, get rid of the excess electrons by combining them with hydrogen and oxygen to form water. Obligate anaerobic microorganisms lack the ability to complete the reaction that leads to the formation of water. In the presence of oxygen, these organisms produce intermediate compounds that are toxic to them. Hence they must avoid oxygen and find other chemicals to act as electron acceptors, such as nitrate (NO_3^{2+}) and sulfate (SO_4^{2+}).

Louis Pasteur on Air

Louis Pasteur was the first to recognize the ability of some microorganisms to grow in the complete absence of oxygen, and in fact he coined the terms aerobe and anaerobe. He showed that controlling the presence of air is critical for making wine, otherwise the ethanol that is formed by the yeast is further oxidized by aerobic bacteria present in the wine to form acetic acid, producing vinegar.

When a batch of wine is begun, the juice is initially aerated to stimulate the growth of the facultative anaerobic yeast. But then conditions soon become anaerobic as the yeast use up the oxygen. The yeast continue in their noble mission while any vinegar bacteria or other spoilers present are inhibited by the lack of oxygen.

Clearly no special procedures are needed to grow aerobes, but obligate anaerobes must be protected from air. This can be accomplished by growing them in sealed tubes in media from which air has been removed. Alternatively, anaerobes may be grown in air-tight vessels called anaerobic jars or anaerobic incubators in which the air has been replaced with an inert gas, usually nitrogen or carbon dioxide.

Growing Viruses

Since viruses are obligate intracellular parasites, to grow them in the laboratory one must be able to grow the viruses' natural host cells. In propagating bacterial viruses (known as **bacteriophages**), one need only grow the proper bacteria, but to grow human viruses, for example, one must provide live human cells in which to culture the viruses. This can be accomplished by techniques known as **tissue** or **cell culture**, where human cells are cultivated in glass or plastic bottles or dishes. Sometimes cells from animals closely related to humans will work, like monkey cells. Fertile chicken eggs also can act as host for some human viruses, such as influenza and yellow fever viruses. The cells or eggs are infected with a virus, then incubated for a period of time to allow the usual viral life cycle to occur. Within a few days the virus population in the culture may have increased by as much as 10,000-fold. The viruses are then harvested and used for further study, or for the manufacture of a vaccine.

Mathematics of Microbial Growth

Many microorganisms reproduce by a process known as **binary fission**. This simply means the organisms divide in two. Thus one bacterial cell becomes two, then four, eight, sixteen, and so on. The remarkable thing is that some organisms are able to reproduce so rapidly that their numbers become astronomical in a very short period of time. For example, some bacteria are able to divide every 20 minutes, in which case one cell will have become over 260,000 cells in just six hours, and over 6×10^{10} cells in 12 hours. Such growth cannot go on indefinitely, however. Depletion of nutrients and accumulation of toxic waste products eventually bring growth to a halt. Otherwise the Earth would be completely covered by a thick blanket of bacteria.

Microbe Math

To calculate how big a bacterial population can become, use the formula

$$Pf = Ps \times 2^g$$

Pf is the final population, Ps is the starting population and g is the number of generations or divisions the population has experienced. Thus if the bacteria divide every 20 minutes, that's three generations per hour. Assuming continuous growth, one cell would become over 1×10^9 in 10 hours (30 generations).

Summing Up

"A single year's life of a bacterium is equivalent to the entire span of mammalian evolution," -Joshua Lederberg, Nobel Laureate, pioneer bacterial geneticist.

Further explorations:

Atlas, Ronald M. Ed. <u>Many Faces Many Microbes</u>. Washington, D.C.: ASM Press, 2000.

Beck, Raymond W. <u>A Chronology of Microbiology in Historical Context</u>. Washington, D.C.: ASM Press, 2000.

Biddle, W. <u>A Field Guide to Germs</u>. 2nd Ed. New York: Anchor Books, 2002.

Blinderman, C. <u>Biolexicon, a Guide to the Language Of Biology</u>. Springfield, IL: Charles C Thomas, Publisher, 1990.

Dixon, Bernard. <u>Unseen Power</u>. New York: Freeman, 1994.

Madigan, M. T., Martinki, J. M. and Parker, J. <u>Brock Biology of Microorganisms</u>. Upper Saddle River, N.J.: Pearson Education Inc., 2003.

David M. Carlberg, Ph.D.

Microbes.info. 23 Mar. 2003. <http://www.microbes.info>

Murawski, Darlyne A. "Fungi." National Geographic Aug. 2000: 58-72.

Prescott, L., Harley, J. and Kline, D. Microbiology. 5th edition. Boston: WCB McGraw-Hill, 2003.

Rosebury, Theodor. Life on Man. London: Paladin, 1972.

Rossmoore, Harold. The Microbes, Our Unseen Friends. Detroit: Wayne State University Press, 1976.

Schlossberg, David, ed. Infections of Leisure. Washington, D.C.: ASM Press, 1999.

Chapter 2

Fire: Controlling Microorganisms with Fire and Ice

The Need for Control

Fire produces heat, and heat is one of the most effective ways of controlling microorganisms. Other methods of control are low temperature (the opposite of heat), chemicals, various types of radiation and filtration.

Why is it necessary to control microorganisms? As we noted in Chapter 1, some of the millions of kinds of microorganisms that share Earth with us can make us sick, spoil our food, and when we do not want them to, cause wood, fabric and other materials to rot. Unless we do something about it, the microorganisms "will have the last word", as the great French chemist and founder of modern microbiology Louis Pasteur once commented.

Terminology

Microbiologists use a number of terms that relate to the control of microorganisms. Here are the most important ones:

Disinfection refers to the reduction in numbers of pathogens present in or on an inanimate object. **Pathogen** means an organism that can cause disease. A **disinfectant** is an agent that is used in disinfection. Bacterial spores, because of their enormous resistance to strong chemicals, are usually not affected by most common disinfectants.

Antiseptics are chemicals that kill or inhibit pathogens and are safe to apply to the body. Like disinfectants, antiseptics have negligible effect on bacterial spores, but are also less effective against vegetative cells than are disinfectants.

Decontamination refers to the reduction of the microbial population in or on an object to some lower value but not necessarily to zero.

Sanitization is the reduction of microbial populations to levels considered safe by public health standards. Restaurants and food manufacturers use **sanitizers** on their food handling equipment. **Sanitizers** are chemical agents that are used sanitize objects.

Sterilization means the complete removal or destruction of viable organisms, including bacterial spores and viruses. Sterility is an absolute condition; an object is either sterile or it is not, whereas an object that has undergone disinfection or decontamination may still contain viable microorganisms. Obviously sterilization requires much harsher conditions than decontamination and frequently a material or object cannot be sterilized easily because the conditions to do so would alter some important quality. Examples are alteration of taste when we attempt to sterilize foods with heat, or the reduction of activity when we use the same method on antibiotics or vitamins. Other methods, discussed below, must be used to sterilize these kinds of sensitive materials. **Sterilants** are chemicals that are strong enough to sterilize objects.

Viable means the ability of an organism to reproduce when placed in a suitable environment. Conversely, **nonviable** is defined as the lack of the ability of a cell to reproduce when it is placed in an environment that would normally support its growth.

Using Heat To Control Microorganisms

Every microorganism has an optimum temperature at which it grows best. If we raise the temperature of its environment above its optimum, growth will be inhibited, and the higher we go, the slower it will grow. Eventually we will find a temperature above which the organism will not grow at all. A common application of this can be found in restaurant buffet tables. The primary purpose of keeping hot foods hot is not to satisfy the customers (clearly they would complain if the food were not hot), but to prevent pathogens and spoilage organisms in the food from multiplying as the food remains on the buffet table. In fact, Public Health laws require that hot foods be maintained at a temperature not less than 140°F (60°C), a temperature at or above which all common pathogens and most spoilage organisms cannot grow. Public health inspectors often carry small thermometers with them which they can insert into the lasagna or the sweet and sour pork at the buffet table to check compliance with the law.

As microbial cells are exposed to yet higher temperatures, critical components such as proteins begin to suffer irreversible damage and the cells eventually become nonviable. **Pasteurization** is a method of exposing certain foods to moderately high temperatures to reduce levels of pathogens and spoilage organisms. Milk is normally pasteurized by heating it to 162°F (72°C) for 15 seconds to destroy pathogens typically known to occur in raw milk. Since some types of fruit juices and other foods occasionally have been found to contain pathogens, they also are frequently pasteurized. Commercially manufactured food items such as sauces, cake batters and cookie dough often contain raw eggs or egg products. Because *Salmonella,* the human intestinal pathogenic bacteria, have been known to occur in fresh eggs, many food manufacturers are now using pasteurized eggs in their products. Beverages that do not normally carry pathogens, such as beer, are also pasteurized primarily to control spoilage organisms.

There are certain species of bacteria, known as thermophiles, that have high optimum growth temperatures, some as high as 110°C. How they are able to survive at high temperatures is not known completely, but it appears that they have evolved proteins and other cell components that are relatively heat-resistant. For example, the cell membranes of thermophiles consist of lipids (fats) that have a higher melting point than those found in mesophiles and are thus able to withstand higher temperatures. Fortunately there are no thermophilic human pathogens, so normal processes to control microorganisms that involve heat are usually effective in protecting us from all common pathogens.

As one might expect, exposing a microbial population to higher temperatures results in a more rapid inactivation of the population. However, inactivation does not occur instantaneously unless very high temperatures are used. Normally the effect of heat on a microbial population acts gradually so it may be some time before the viable population is reduced to zero. Figure 2.1 illustrates this important concept.

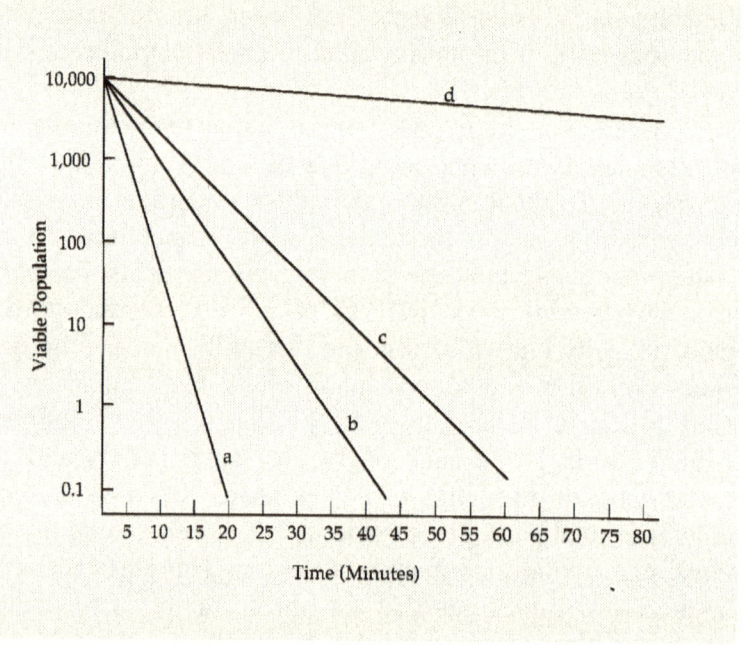

Figure 2.1. Graph plotting the death of vegetative forms of three species of bacteria

and bacterial endospores when exposed to a temperature of 100°C (212°F), the boiling point of water. Populations are plotted on the vertical axis, time of exposure on the horizontal axis. Of the three vegetative forms shown (a, b and c), a is the more sensitive to heat, c is the least sensitive. Bacterial endospores (d) are the most resistant form, showing almost no effect of the heat.

The steepness of the lines in Figure 2.1 is a measure of the relative susceptibility a given population of microorganism shows toward heat. Microbes vary in their sensitivity toward heat. For instance, hepatitis B virus is famous for being relatively heat resistant compared to most other viruses, while the bacterium *E. coli* shows about average heat susceptibility. Bacterial spores are the champions of heat resistance. They can withstand boiling water for extended periods of time, which means considerably higher temperatures are needed to kill bacterial spores. This brings us to the subject of sterilization with heat.

Sterilization, as you recall, refers to the complete removal or destruction of microorganisms in or on an object. Heat is normally used to sterilize objects such as bacteriological media and surgical instruments. One of the most critical factors in using heat to sterilize objects is the presence of ambient moisture. The greater the amount of moisture that is present, the more effective is the heat. Thus the most effective level of moisture is 100% relative humidity (RH). To illustrate this, a population of bacterial spores, which are notorious for their heat resistance, can be reliably killed in 20 minutes at 121°C (250°F) in the presence of 100% RH. It would take several hours at the same temperature to kill them at 50% RH.

The primary reason for the difference in effectiveness between moist and dry heat is due to the mechanism that damages cell proteins and leads to cell death. Saturated moisture in the presence of heat causes cell proteins to congeal and thus become inactivated, which is a fairly rapid reaction. The same thing happens when an egg is hard-boiled. In addition, water is a better conductor of heat than dry air. In the absence of significant moisture the principal reaction that kills the cells involves a different mechanism, the inactivation of critical molecules by oxidation, a relatively slow chemical reaction. Because the presence of moisture in sterilization is so critical, microbiologists distinguish two types of heat. Heat applied in the presence of 100% RH is referred to as moist heat. Heat that is accompanied with anything less than 100% RH is considered dry heat.

Moist heat sterilization

Moist heat sterilization, which combines heat with water, is normally carried out with steam under pressure in an apparatus known

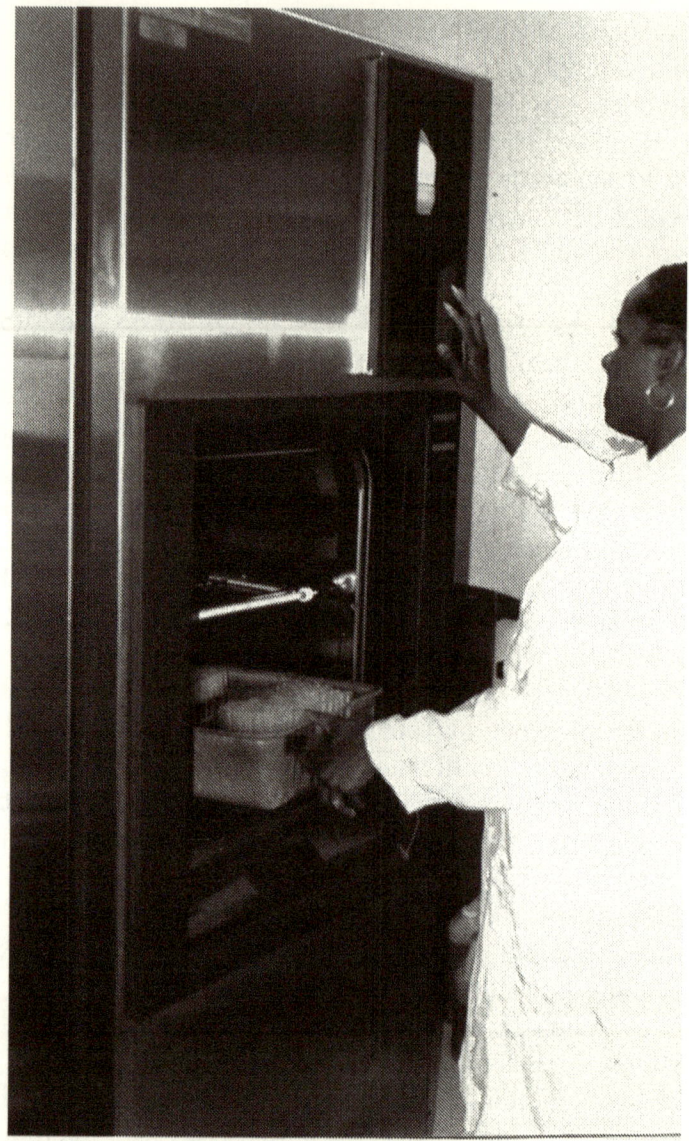

Figure 2.2. An autoclave for sterilizing items with moist heat supplied by steam under pressure.

as an **autoclave** (Figure 2.2). Steam under pressure is needed for three reasons: to achieve 100% RH, to reach a temperature high enough to kill bacterial spores in a reasonable period of time and to prevent aqueous liquids from boiling over. The standard temperature for sterilizing objects in an autoclave is 121°C which is produced with steam at a pressure of 15 PSI. Twenty minutes is usually sufficient time to sterilize small objects such as surgical instruments and small volumes (less than a liter) of salt solutions or bacteriological media. Large objects require proportionally longer times. Occasionally higher temperatures and shorter times can be used if the materials can withstand the harsher conditions.

Bacterial Spores

Bacterial spores are extremely resistant to high temperatures, requiring heat considerably higher and applied for longer periods of time to kill them than would be necessary to kill the vegetative forms of the same bacteria. Three spore-forming species of bacteria can cause serious diseases: botulism, gas gangrene and tetanus. Botulism, a relatively rare form of food poisoning that is often fatal, occurs when the botulism bacteria (*Clostridium botulinum*) grow in food and release a potent neurotoxin. When ingested, the toxin can produce the disease known as botulism. Most cases of botulism can be traced to eating food that had been improperly canned. To avoid botulism, any food that may contain botulism spores must be processed with sufficient heat during canning to kill the spores. In the home, only a pressure cooker can provide the proper heat. Since botulism bacteria are commonly found in soil, fresh produce is the most likely food to contain botulism spores. Ingesting *C. botulinum* spores directly is harmless to normal, healthy adults, however.

Gas gangrene and tetanus are caused by the spore-forming bacteria *Clostridium perfringens* and *Clostridium tetani*, respectively. These organisms also are very common in soil. Any object that has dust or soil on it may contain the spores of these organisms. To avoid these serious infections, any items that are intentionally used to penetrate the skin such as surgical instruments, hypodermic needles, and tattoo and skin-piercing needles must be sterilized with heat or other means sufficient to kill bacterial spores. Accidental puncture wounds caused in automobile accidents or by nails and rose thorns have been some of the more common ways of contracting tetanus. The familiar story of the danger of stepping on a *rusty* nail is based on the assumption that a rusty nail has been in the ground longer and is more likely to be contaminated with tetanus spores than a new, shiny nail. A very effective vaccine for preventing tetanus is widely available. *Clostridium perfringens* also causes a common type of food poisoning.

Dry heat sterilization

Dry heat sterilization must be carried out at considerably higher temperatures and for longer periods than what is required in moist heat sterilization. Two or more hours at 320°F–356°F (160° to 180°C) is normally used to sterilize objects with dry heat. You might ask why dry heat is even considered for sterilization if it is so much less effective than moist heat. In certain situations, dry heat may be the preferred method. For example, to sterilize sealed, empty glass bottles in an autoclave would not be possible because the steam (moisture) would not reach the interior of the bottles. Empty bottles can only be sterilized in an autoclave if the caps are loosened or if the bottles are closed with cotton plugs to allow entry of the moisture. Nonwettable materials such as talcum powder, sand, mineral oil and petroleum jelly cannot be sterilized in an autoclave because the steam will not penetrate them. Dry heat is the best alternative for such materials as clean, empty laboratory ware like glass bottles, Petri dishes and pipettes. Liquids that boil below 180°C, most fabrics and plastics and heat sensitive chemicals cannot be sterilized with dry heat. Dry heat sterilization can be carried out in any oven-like apparatus that can reach 160° to 180°C.

How Sterile is Sterile?

While sterility is considered an absolute condition, sterilization is not. When do the items that are being sterilized become sterile? Clearly when the curve (Figure 2.1) drops below a population of 1 cell (you can't have 1/10 of a cell), that must mean sterilization has been accomplished. In practice when the scale on the vertical axis of our graph drops below 1, it changes from showing the viable population to showing a probability that the items remain unsterile. That is, if we sterilize our items until the curve reaches 0.01 (1/100) on the vertical axis, it means there is still a 1 in 100 chance some of our items are not sterile. Thus when sterilizing a batch of 1000 bottles of saline for injection, statistically 10 bottles may not be sterile (0.01 X 1000). The Food and Drug Administration requires

---

that critical items such as saline for injection and other materials that come in direct contact with the blood stream must be sterilized to a probability of 10^{-6} (1/1,000,000), referred to as the **sterility assurance level**. That sets the chances of a patient being exposed to a non-sterile item at one in a million.

Using Low Temperature To Control Microorganisms

Referring back to Figure 1.8 in Chapter 1, one can see if the temperature is lowered from the optimum, the growth of the bacteria also is affected. Thus **refrigeration** retards food spoilage by inhibiting the growth and activities of many spoilage microorganisms. But as you recall, psychrophilic and psychrotrophic microorganisms can grow at temperatures below 20°C and some can grow at the temperature of a home refrigerator, about 4°C. That is why food stored in the refrigerator still eventually spoils. Human pathogens are mesophilic and with a few exceptions cannot grow well at refrigerator temperatures. Public health laws require that foods in restaurants such as salad dressings and custards that are often associated with food infections or poisoning must be kept at a temperature under 5°C (41°F).

Freezing, that is, storage below 0°C, is most effective if carried out below −10°C. Home freezers operate at about −20°C (−4°F), which effectively inhibits microbial activity. Unfortunately other types of non-biological food spoilage can still occur in home freezers, such as dehydration ("freezer burn"), oxidation (darkening) and rancidity. Freezers for commercial and scientific uses operate as low as −70°C (−94°F) where nearly all types of spoilage are averted.

The inhibition of microbial growth by low temperatures is largely reversible. For example, if you return some frozen food to room temperature, many of the microorganisms on it will quickly start to grow. Freezing is one way of preserving microbial cultures in the laboratory for future use. Some types of frozen food that have been thawed are frequently more susceptible to spoilage than when fresh. This is because freezing alters the cellular structure of the food, making it more readily attacked by microorganisms. That is why grocery stores must alert consumers when foods have been frozen and then thawed.

Chemical Methods of Control

It is well known that many chemicals are toxic to living cells. Examples include organic compounds such as phenol (carbolic acid) and ethyl alcohol and high concentrations of metals like mercury, copper and lead. Microorganisms are highly susceptible to these chemicals and therefore the chemicals can be used to control the activities of the organisms. However, since many of these chemicals are also hazardous to humans and other animals, they must be used with care.

Names of classes of chemicals that kill microorganisms frequently end with -cide, with the root of the word referring to the kind of organism affected. That is necessary because certain chemicals may be effective toward one type of organism but not others. Products that kill bacteria (but not spores), fungi, or viruses are called **bactericides**, **fungicides**, or **viricides**, respectively. Considerable overlapping occurs. For instance, many bactericides may exhibit fungicidal and viricidal activity as well. Products that are sufficiently powerful to kill bacterial spores are called **sporicides**. If a product only prevents growth of the organisms but does not kill them, the suffix -stat is used: **bacteristat**, **fungistat**, and so forth. Fungistats are often added to paints to prevent mildew. Some chemicals are -cides at high concentrations but become -stats at lower concentrations. Since exposure to -stats does not kill microorganisms, growth normally resumes when an agent is removed.

The United States Food and Drug Administration and the EPA are responsible for monitoring products that claim to have antimicrobial activity. Through investigations by these agencies, many companies have been assessed fines, had their products seized, and were required to alter their advertising and labels when their products did not live up to their claims. Typical household products that have been targets of federal prosecution due to unsupported claims have been mouth washes, bathroom disinfectants and toys. Products sold to doctors and hospitals also have been under federal action for unsupported claims. Clearly one must remain skeptical whenever a manufacturer claims extraordinary antimicrobial activity for a product.

Here are a few general examples of chemicals that are used to control microorganisms:

Disinfectants, Antiseptics and Sanitizers

Many chemical agents are not sufficiently potent or reliable to act as sterilants but are still useful in the control of microorganisms. These are the disinfectants, antiseptics and sanitizers.

Quaternary ammonium compounds ("quats") are popular disinfecting agents. Examples are benzalkonium chloride and cetyl pyridinium chloride. Since they are actually detergents, some of their antimicrobial activity is due to their ability to physically remove microbes from the skin and inanimate objects. However, they are also moderately effective bactericides in their own right, and at concentrations as low as 0.0005 percent they can still be bacteristatic. Most popular mouthwashes contain these agents, which underscores their relatively low toxicity for humans. Quats cannot be relied upon to disinfect medical instruments and similar critical applications since they are not effective toward certain types of bacterial pathogens such as *Pseudomonas,* tuberculosis bacteria and bacterial spores.

Phenol (carbolic acid) is an effective disinfectant, but it is highly toxic and corrosive to the skin, which severely limits its general use. Derivatives of phenol, known as **phenolics**, such as orthophenylphenol, are considerably less toxic and actually more effective than phenol. Phenolics are used in many common household disinfectants. The phenolic compound known as hexachlorophene is a good skin antiseptic, but its use must be stringently controlled. It can be absorbed through the skin and is highly damaging to the central nervous system of newborn babies, though it is relatively safe for use by adults. Another type of phenolic, triclosan, has become a common additive to many household products, including soaps, lotions, toys, clothing and kitchen cutting boards.

Alcohols, such as ethanol (grain alcohol) and isopropanol (rubbing alcohol), are effective and widely used to disinfect skin prior to inoculations or taking of blood samples. As in the case of the quats, the effectiveness of the alcohols appears to be a combination of cleansing action and chemical activity. Water solutions of about 70–90 percent (by volume) of ethanol or isopropanol are bactericidal against vegetative cells but have little effect on bacterial spores. Because they have a tendency to evaporate quickly, the effectiveness of alcohols is frequently reduced unless they are applied generously.

Chlorine and iodine belong to a group of chemical elements known as halogens. Chlorine is a gas at room temperature and is commonly used as a disinfectant of drinking water and waste water. Chlorine dioxide, also a gas, and organic chlorine compounds have frequently replaced chlorine as a disinfectant in many applications because they are safer to handle. Inorganic chlorine compounds, principally hypochlorites (ordinary laundry bleach), also are widely used as disinfectants and sanitizers in water and waste water treatment and in the food industry, including dairies and restaurants. They are powerful oxidizers and highly microbicidal.

Chlorine gas is the most common chemical for disinfecting water, but organic matter in the water tends to neutralize the chlorine and its compounds. If water has an excessively high organic content, such as waste water or heavily polluted water, the amount of chlorine to be used must be greatly increased. However, increased chlorine concentrations in the presence of organic matter trigger the formation of chlorinated hydrocarbons, many of which are carcinogens. For that reason cities such as New Orleans, which depends on the Mississippi River for much of its drinking water, have stopped using chlorine for disinfecting their water. Water that has been heavily used to dispose of waste, like the Mississippi River, will contain high concentrations of organic chemicals that can become chlorinated. To avoid the formation of chlorinated hydrocarbons, cities have switched from chlorine gas to chloramines or chlorine dioxide. **Chloramines** are organic chlorine compounds that do not form chlorinated hydrocarbons, but they are not as effective as chlorine gas.

The problems that cities such as New Orleans experienced with the formation of hazardous chlorination by-products led to amendments to the federal Safe Drinking Water Act (SDWA) in 1996. Included in the amendments are the Microbial and Disinfection Byproduct Rules, which limit the amount of chlorinated byproducts that can be in drinking water. The rules recognized two somewhat conflicting principles which public water systems must find ways to implement simultaneously:

•Sufficient disinfectant must be used in drinking water to inactivate pathogens

•Disinfectant residuals and byproducts must be maintained below harmful levels

The EPA guidance manual states, "Public water systems will have to make complicated technological and economic decisions to comply with the different drinking water regulations."

Chlorine dioxide (ClO_2) is a gas like chlorine and has been used as a substitute for chlorine to decontaminate water in cooling towers and potable water systems. ClO_2 has several advantages over chlorine. It is not as susceptible to inactivation by organic matter and does not form significant amounts of potential carcinogens. It is safer to handle than chlorine gas because it is usually generated on site. That is, an amount of gas is produced just sufficient for a given operation, eliminating the hazards of transporting and storing large amounts. ClO_2 received much press coverage when it was used to decontaminate offices following the recent terrorist delivery of anthrax spores through the mail.

Iodine has been a familiar household antiseptic for generations, usually as an alcoholic solution known as a **tincture**. **Iodophors**, organic compounds of iodine, have widely replaced tincture of iodine for the treatment of wounds, for they are somewhat less irritating to open tissue and have thus gained the nickname "ouchless" iodine. Iodine and its compounds are good skin antiseptics, because they bactericidal and fungicidal, and are widely used on skin before surgical procedures. Other uses for iodine and iodophors include disinfection of water and food handling equipment. Examples of these products are Wescodyne® and Isodine®.

Hydrogen peroxide has long been a popular antiseptic and disinfectant. Like chlorine and its compounds, it is a strong oxidizer. Used in concentrations of three to six percent, it is a reliable disinfectant *on clean objects*; it is unstable in the presence of dirt and organic debris. Higher concentrations are sporicidal. It is also used as an alternative to chlorine disinfection for waste water and as a vapor sterilant in certain industrial applications.

Ozone (O_3) is considered the most effective drinking water disinfectant, but because it does not leave a residual like chlorine does, its use is generally limited. A residual means that a disinfectant remains active in the delivery system as the water is piped to homes and businesses. Although it produces an unpleasant taste and odor to drinking water, residual chlorine continues to control microbial populations right up to the consumer's tap. When the use of ozone is necessary to assure a higher level of disinfection but a residual is desirable, ozonation is usually followed by chlorination.

Certain metals, such as **mercury, silver, zinc and copper**, are inhibitory toward many microorganisms. Mercury bichloride has been widely used as a wood and anatomical specimen preservative and a antiseptic and disinfectant. Silver and zinc are bacteristatic and have been used as antiseptics for many years. Examples are silver nitrate, silver oxide, "Argyrol" (an organic silver product), and zinc oxide. Copper is often added to water to prevent the growth of algae. Ironically, while some metals like zinc and copper are toxic at high concentrations, they are actually necessary nutrients at low concentrations, referred to as "trace elements".

Food Preservatives

As we have mentioned on several occasions, the activities of some microorganisms cause food spoilage. Spoilage is defined as a reduction in the organoleptic quality of a food, which means any quality that can be measured with the senses: taste, appearance, feel or smell. Food products pick up spoilage organisms from a number of sources: during their harvesting, processing or shipping, in the warehouse, in the retail store and in the consumer's kitchen. If conditions are right, the microorganisms will grow in the food and release by-products such as unpleasant tasting acids, foul smelling gases and amines, ugly colored pigments and slimy carbohydrates, the presence of any of which will cause one instantly to reject the food. Not all food spoilage is caused by microorganisms, however. Pure chemical changes such as oxidation (the darkening of apples, lettuce and potatoes, for example) also frequently occur. While spoiled food represents an economic loss for manufacturers, merchants and consumers and may occasionally pose a health hazard if the food is eaten, it also presents a serious environmental problem in that its disposal adds additional volume and organic load to our waste water or solid waste disposal systems. These are all strong incentives to control food spoilage.

Examples of methods for reducing spoilage are the use of sturdy packaging, refrigeration, and the addition of chemicals to the food that inhibit or kill the spoilage organisms as well as prevent chemical spoilage. Obviously since food is ingested, the added chemicals, generally known as **preservatives**, must be very carefully selected. Chemical preservatives occur in foods in three ways: they are natural ingredients, they are produced in the food during processing, or they are added to the food. Benzoic acid, diacetyl and sodium nitrite are three examples of common preservatives. Benzoic acid is a natural preservative that is abundant in certain fruits such as cranberries, diacetyl is formed during the fermentation of milk to make butter, buttermilk and yogurt, and sodium nitrite is added to cured meats to inhibit the growth of botulism bacteria. Most chemical food preservatives are effective toward specific types of microorganisms such as molds or bacteria. Potassium sorbate, primarily active against molds, is added to jams, jellies, salad dressings and fruit juices because these foods are particularly susceptible to mold growth; they are too acid to support the growth of most bacteria.

The first preservative added to food was probably salt, produced by the evaporation of water from saline seas and lakes, or extracted from mines. In these earliest uses of a preservative, which go back several thousand years, the salt was probably a crude mixture of sodium and potassium chloride, sodium sulfate and other salts. Modern day "table salt" is essentially pure sodium chloride (NaCl). Salt draws water out of the cells of spoilage organisms through osmotic pressure, eventually inhibiting their activities. Sugar acts the same way, which is why strawberry jam and maple syrup do not spoil readily.

Over the centuries a number of substances have been used to preserve food, some of which are still used, while others have been abandoned when new information revealed their potential toxicity. Before governmental controls were adopted in 1906, United States food manufacturers added such chemicals as borax and formaldehyde to foods to prevent spoilage. All food additives now must be thoroughly tested and approved by the United States Food and Drug

63

Administration (FDA) before they can appear in a product. There is, however, a so-called **GRAS** (Generally Recognized As Safe) list of approved chemicals that have been used for many years without any evidence of detrimental effects when consumed. They do not have to be re-tested when proposed for a new food product.

Many foods such as milk, fruits, fruit juices and vegetables may be subjected to processes that convert them into products that spoil less readily. When milk is changed into cheese or cabbage into sauerkraut, their keeping quality is greatly extended. In a process generally known as **fermentation**, an immense variety of perishable foods become not only less susceptible to spoilage, but gustatorily more interesting. The making of some fermented foods, such as cheese, is at least 9000 years old. There are thousands of different types of fermented foods that are produced worldwide and in most cases they define their ethnic origins. More familiar examples are yogurt, pickles, sauerkraut, kim chee, soy sauce, fish sauce, poi and wine. During fermentation, microorganisms in the food (either added or already in it) convert sugars and other constituents into natural preservatives such as acids and alcohols, creating conditions that are inhibitory to most spoilage organisms. Adding salt and reducing moisture also are frequently involved in the production of fermented foods, resulting in even greater keeping qualities.

Preserving Materials

Natural materials such as wood and paper are common menu items for a number of microorganisms, particularly fungi. Wooden objects that come in contact with soil or water such as fence posts, playground equipment, bridge foundations and telephone poles would have to be replaced fairly often if it were not for the various chemical preservatives that can be applied to the wood. Chemicals containing arsenic or copper are often used to protect some common wooden items from rotting. These chemicals are usually too expensive for use in large amounts, such as on telephone poles or bridge pilings. Instead, the familiar black coal tar creosote is usually used. Creosote contains phenolics, derivatives of phenol which are excellent antimicrobial compounds. Mercury has been added to latex paints to inhibit mildew, but its use was banned by the EPA in 1991 when it was shown that mercury vapor was being released into homes and businesses.

Sterilants

Sterilants are chemicals that are strong enough to sterilize objects. There are only a few reliable chemical sterilants.

The three most popular chemical sterilants are **ethylene oxide** (ETO), **formaldehyde**, and **glutaraldehyde.** ETO is a gas that is commonly used to sterilize medical supplies as well as non-medical products such as imported spices, cosmetics, and museum artifacts. In spite of its flammability, toxicity and potential carcinogenicity, about 45 percent of all medical supplies manufactured in the United States is sterilized with ETO.

Formaldehyde is a gas that has been employed as a sterilant for over 100 years, but it has a very irritating odor that makes it unpleasant to use. It is also a suspected carcinogen. It is sometimes used to sterilize air ducts and rooms that have become contaminated. An aqueous solution of formaldehyde, known as **formalin**, has been used for preserving biological specimens, but it is rapidly being replaced by less hazardous substances.

Dilute solutions of **glutaraldehyde** are frequently used in medical and dental offices to sterilize small surgical devices, catheters, and similar items, but they must be rinsed with sterile water before use to remove residual agent. Glutaraldehyde has the capacity to kill bacterial spores but only after extended exposure.

Factors that Affect the Chemical Control of Microorganisms

Several factors are known to alter the effectiveness of chemical sterilants and disinfectants. Temperature and concentration are parameters that are under the control of the user. As a general rule, elevated temperatures improve the effectiveness of chemical agents provided the agents do not decompose or evaporate. Increasing the concentration of an antimicrobial agent will not necessarily improve its effectiveness. More is better is not always the rule. For example, 70 percent ethanol is more effective as a disinfectant than 100 percent ethanol.

David M. Carlberg, Ph.D.

Radiation

Radiation is defined as electromagnetic energy. Examples of radiation are radio and TV transmissions, light and x-rays. In order for a particular type of radiation to have any impact on microbial cells, it must have sufficient energy to damage the cells. Other than the incidental heat that is generated, the radiation from a microwave oven, for example, is too weak to have any direct effect, but ultraviolet light, x-rays and gamma radiation all have sufficient energy to bring about changes in cells that can lead to their death. While ultraviolet light (at a wavelength of about 260 nanometers) is effective in killing microorganisms, it is not energetic enough to penetrate glass, plastic or dust and sediment. Ultraviolet light is therefore most useful in the disinfection of clean water and air or the surfaces of smooth, clean objects. Any particulate matter present will shield microorganisms from the effects of the UV light and drastically reduce its effectiveness.

More powerful types of radiation, known as ionizing radiation, such as gamma radiation and so-called cathode beams (also called electron beams or e-beams), are penetrating enough to sterilize bulky items such as medical supplies in their final shipping packages. Facilities to produce ionizing radiation for sterilization are very expensive to build and maintain, and are also very hazardous to operate. But for certain types of medical supplies there are no alternate methods, and in fact about half of all medical supplies manufactured in the United States are sterilized by some type of ionizing radiation.

Cobalt-60 (60_{Co}) is a common source of ionizing radiation for the sterilization of medical supplies and other products. This radioactive isotope of cobalt emits very energetic gamma radiation that is capable of penetrating large containers of product. The 60_{Co} is placed in the middle of a large room that has 10 foot thick walls to protect nearby workers. Products to be sterilized are transported into the room on flat-bed cars or a monorail, moved past the 60_{Co} source and then exited. The intensity of radiation that is needed to sterilize an object is about 10,000 times greater than that of an ordinary chest x-ray.

Contrary to many beliefs, objects do not themselves become radioactive when sterilized with radiation.

Radiation of food

Approximately 76,000,000 cases of food-borne illness occur in the United States each year, leading to 325,000 hospitalizations and over 5000 deaths, in all representing enormous social and economic losses. Because ionizing radiation has little or no effect on the quality of foods such as meats and fruit at the intensities commonly used, it seems an ideal way to rid food of pathogenic and spoilage organisms. Certain products such as imported spices, dry seasoning mixes and potatoes have been treated with radiation for many years. Raw spices, because of the manner by which they are harvested and processed, may carry human pathogens, which are inactivated by low doses of radiation. Radiation of potatoes inhibits sprouting. The FDA approved the use of low doses of radiation for reducing microbial contamination in pork in 1985, in raw chicken meat in 1992, and in eggs, meat and meat products in 2000. To avoid health problems during space flights, NASA has been feeding its astronauts irradiated food for many years.

Food manufacturers have not taken full advantage of radiation to reduce the threat of food spoilage and food-borne diseases. There seems to be two reasons for this: cost and the uncertain acceptability by consumers. Irradiated food costs a few cents more per pound to process, an added expense manufacturers fear would not be accepted by consumers. Also, in spite of the threat of food-borne disease and over 50 years of testing that shows irradiated food to be safe, consumers are still cautious about buying irradiated meat and chicken. Consumer doubts were confirmed by a survey conducted in the 1990s in which only about 50 percent of those polled would buy irradiated meat and chicken, and only half of the 50 percent group (25 percent overall) would be willing to pay more for it. With this kind of uncertainty, food manufacturers are not about to invest large sums of money on radiation equipment. Another striking revelation of the consumer survey that less than half of those polled had ever even heard of food irradiation.

David M. Carlberg, Ph.D.

Antibiotics and Antimicrobics

Microbiologists have recognized at least from the time of Louis Pasteur that certain microorganisms, particularly molds and bacteria, produced substances that in exceedingly low concentrations inhibited or killed other microorganisms. These substances are now known as **antibiotics**, and penicillin is no doubt the prime example of such a substance. Discovered in 1929 by the Scottish bacteriologist Alexander Fleming, penicillin proved to be an enormously effective and safe agent for treating a variety of bacterial infections. During World War II, when it was realized that more soldiers were dying of wound infections than from the wounds themselves, the large scale production of penicillin was undertaken. The antibiotic was no doubt responsible for saving innumerable lives. Since the introduction of penicillin, thousands of antibiotics have been discovered (see box below), but only a couple of hundred have shown the necessary characteristics as successful agents for treating infections: high potency against the pathogen with low toxicity for the patient. Organic chemists have found ways of altering many natural antibiotics to improve their properties. These are called **semi-synthetic antibiotics**, and two examples are methcillin and ampicillin, derivatives of penicillin.

The Great Quest

Since most antibiotic producing microorganisms live in soil, over decades the search for new antibiotics prompted the collection of millions of soil samples from all over the world. Whenever a pharmaceutical company employee traveled, he or she was encouraged to carry a supply of plastic vials with which to scoop up soil samples along the way, which were then brought back to be tested for antibiotic producing microbes. Microorganisms that produced promising antibiotics were found in soil samples from countries all over the world. For instance, a soil sample that was reportedly picked up by a tourist visiting Italy's Roman Forum contained bacteria that were found to produce the useful antibiotic, spiramycin.

Totally synthetic drugs for the treatment of infections, generally known as **antimicrobics**, also have a long history. The first successful agent was arsphenamine, an arsenic-containing chemical developed by the German bacteriologist Paul Ehrlich in 1910 for the treatment of syphilis. Sulfonamide was introduced by the German bacteriologist Gerhard Domagk in 1932, followed by a long list of derivatives commonly known simply as "sulfa drugs". These agents have been very effective in treating a variety of bacterial infections.

While most drugs for treating bacterial infections are relatively free of unpleasant side-reactions, drugs for treating serious fungal and protozoan infections often produce complications for the patient. This is because the pathogens and the patient's cells are eukaryotic and thus both often possess identical or similar susceptible targets. Physicians must weigh the side-effects of a drug against the consequences of the infection and sometimes it is better for the patient to omit drug therapy than to risk a serious side reaction. The development of new drugs for treating infections by eukaryotic microorganisms has been slow partly because of the ever-present toxicity problem and partly because of the limited economic incentive in producing such drugs. Serious fungal and protozoan infections are relatively rare in developed (read *wealthy*) countries and relatively common in developing (*not wealthy*) countries.

Viruses are generally not susceptible to the antibiotics and other common chemotherapeutic drugs normally used against other microbial infections. That is because viruses do not have a metabolism of their own or a complex cell chemistry that can be targeted by antimicrobial drugs. As you recall, viruses depend on the host's cells for reproduction. The drugs that are available for the treatment of virus infections aim at those few specific targets that are unique to the infecting virus. Examples are amantidine for influenza, acyclovir for herpes, and AZT for HIV infections. Often drugs for treating virus infections do not cure the infection, but only alleviate symptoms or reduce infectivity.

Drug Resistance

With the development of dozens of powerful antimicrobial drugs during the middle decades of the 20th century, it was assumed infectious disease would be eradicated by the end of the century.

Instead, Nature has provided microorganisms with various mechanisms to acquire resistance toward these drugs. Resistance means a given microorganism has the ability to withstand concentrations of an antimicrobic drug that would normally kill or inhibit it. Nature has not been alone in the development of antibiotic resistance; certain human practices have helped her. Under natural circumstances, most bacteria are susceptible to the majority of antimicrobial drugs. However, there have always been a minority population of resistant microbes lurking in the background. They acquired the resistance through normal genetic mechanisms independent of the presence of the drugs. But over the decades that antimicrobial drugs have been available, in some instances the minority resistant forms have gradually become the majority. What has brought this about? Our environment has become saturated with antibiotics. About half of all antibiotics manufactured today are fed to farm animals, such as chickens, turkeys, hogs and beef cattle. That is because it was discovered some years ago that animals grew faster and healthier when given antibiotics as part of their normal diet. Farms have become breeding grounds for antibiotic-resistant bacteria.

In addition, some physicians have long over-prescribed antibiotics for their patients. Many patients feel they are better cared for if they leave their doctor's office with a prescription for an antibiotic in hand, regardless of the nature of their illness. The end result is that the low levels of antibiotics that have built up in our surroundings tend to inhibit the susceptible microorganisms, but leave the few resistant ones alone. This means that over time there is a shift from antibiotic susceptible populations to resistant ones. Then, if one develops a bacterial infection, there is an increasing chance the pathogen will be resistant to the antibiotic that would normally be used. By the time a physician discovers the pathogen is resistant to the antibiotic that was prescribed, the infection may have progressed to a serious stage and the physician must call in a clinical microbiologist to determine what drug would be appropriate for the particular pathogen. The situation has gotten so critical that in spite of years of searching for more powerful drugs, there are simply no effective antibiotics to treat certain types of infections; the organisms are resistant to everything that is available.

Controlling Microorganisms By Filtration

There are some materials that cannot be sterilized by heat, chemicals or radiation due to their sensitive nature or the expense. Examples are vitamins and other drug solutions and certain products like those for the care of contact lenses, which contain heat-sensitive enzymes and other substances. Filters made of various polymeric substances have been developed that have a porosity fine enough to trap bacteria and other microorganisms. Sterilization is accomplished by passing liquids through these materials, known as **membrane filters**. Similar filters are used to "cold pasteurize" liquids like beer and fruit juices to remove spoilage organisms. These same filters are used to determine bacterial populations in water. That is covered in Chapter 3.

Summing Up

Because we are in constant contact with enormous numbers of microorganisms throughout our daily activities, we have had to devise various physical and chemical methods to keep the harmful ones under control without seriously affecting the beneficial ones or ourselves. Our success has been tenuous. Infectious diseases continue to be major causes of debilitation and death in developed and developing countries alike, and food spoilage and material deterioration persist.

Further explorations:

Nicolaou, K. C. and Boddy, C. N. C. "Behind Enemy Lines." Scientific American May 2002: 54-61.
Radetsky, Peter. "Last Days of the Wonder Drugs." Discover Nov. 1998:76-83.

Chapter 3

Water: The Essential Substance

The Qualities of Water

A few years ago, SCIENCE, the weekly publication of the American Association for the Advancement of Science, each year announced a Molecule of the Year award for the chemical that had made the greatest impact on history that year. If a Molecule of All Time were ever selected, it would have to be water. Water, making up 70 to 90 percent by weight of all living cells, is absolutely essential for life.

Water possesses a number of physical and chemical properties that make it a unique and useful substance. The water molecule (H_2O) is electrically **polar**, that is, one part is negatively charged and one part is positively charged. That makes it compatible with lots of different kinds of molecules and thus it is an ideal solvent that is able to dissolve the wide variety of biochemicals that are the necessary components of living cells. Water has a relatively high **specific heat**, meaning it takes a lot of energy to heat it up to a particular temperature, but then it can hold that temperature over a long period of time. That means the temperatures of aquatic environments from oceans and lakes down to the cytoplasm of a single bacterial cell will remain relatively stable in spite of brief temperature fluctuations in the surrounding environment.

Rain drops form in clouds because water molecules like to stick together, a characteristic known as **cohesiveness**. Water molecules also like to stick to other materials (except a duck's back), making the materials wet. That feature is called **adhesiveness** and provides, for example, protective measures for our bodies, such as preventing membrane surfaces (eyes, respiratory passages) from drying out. The combination of these two attributes, cohesiveness and adhesiveness, is what gives water the ability to exhibit **capillary action**, the force that helps raise water and nutrients to the tops of even the highest trees, the giant redwoods.

Another feature of water is attractiveness, especialy when the water is unpolluted and held in large bodies such as lakes or oceans. It is a quality that scientists have not been able to explain. Evidence of this quality is found world-wide: the cost of water-front property.

Water, that simple substance, also plays a critical role in many day-to-day functions of our modern society. Here is a look at how we use that special liquid we take for granted.

The Uses of Water

Domestic and public uses
    Drinking
    Cooking
    Bathing
    Laundering
    Irrigating landscape
    Removing waste
    Washing cars, pets, sidewalks, etc.
    Esthetics (fountains, lakes and other water features)
    Recreation (swimming, boating, water skiing, fishing)
Industrial uses
    Production
        Direct (foods, beverages)
        Indirect (paper, coal, steel, printed circuits)
    Cooling
    Hydroelectric power generation
    Steam power generation
    Removing waste
Agricultural uses
    Irrigating crops

> Washing harvest
> Watering stock animals
> Aquaculture

Since seventy eight percent of Earth's surface area is covered with liquid water, scientists often refer to our home as The Water Planet, a term said to have been coined by Earth's first aquanaut, Jacques-Yves Cousteau. Of all the water on Earth, ninety seven percent is in the seas and oceans and is generally not directly usable for many of the applications listed above due to its salt content, or is inaccessible due to location. A little over two percent of Earth's water supply is stored as ice in our polar caps, also inaccessible and essentially unusable. Therefore, less than one percent of Earth's water supply is liquid, fresh water and directly usable by human populations. In the United States, water consumption for domestic use varies from about 100 to 300 gallons per day per person depending on the region. Factor in industrial and agricultural uses and the daily usage is over 1000 gallons per person. On that basis, the American population uses over 280 billion gallons of water every day of the year.

Conserving Water

The total water supply of Earth has not changed significantly since the planet's beginnings. With Earth's rapidly growing human population, it is entering a period when in many parts of the globe there simply will not be enough usable, unpolluted water to satisfy local needs. Presently, over a billion people lack running water in their homes and awake each morning not knowing where they will find water to fill that day's needs. In a 2001 report, the United Nations released a study in which it was predicted that by 2025, one third of the world's population will experience serious water shortages. This will drastically affect food supplies, since agriculture depends on adequate supplies of water and accounts for 70 percent or more of water use. If our present living standard is to continue, we must consider possible solutions to the water problem. Here are three:
1. Practice more efficient water use

2. Develop new sources of usable water
3. Recycle the water we use

By encouraging (and in some areas requiring) limits on water use, some success has been achieved in controlling water consumption in the United States. From 1950 to 1980, water consumption outpaced the nation's population growth rate. Water consumption peaked in 1980, then dropped more than ten percent in spite of a 10 percent increase in population, and then remained nearly level through 1995 (the last year figures are available). Examples of water conservation strategies are: more efficient management of agricultural water, the use of low flow shower heads and toilets, intensive inspection programs to locate leaking plumbing, and the introduction of educational programs to acquaint citizens about the problem of future water shortages. For example, with its water conservation program that began in 1994, the city of New York estimates it saves nearly 100 million gallons of water a day. When Mexico City replaced 350,000 leaky toilets, the water saved could supply water to a quarter of a million new residents.

The oceans are an obvious alternate source of water, but the cost involved in the desalination of sea water compared with other water sources has discouraged its general use. In certain arid regions of the world, such as the Middle East or isolated islands, where there is little or no natural fresh water, the cost of sea water desalination for producing drinking water has been readily accepted. Presently, desalination accounts for only about one percent of the world's source of drinking water, but that may be changing. Thanks to improved technology introduced in recent years, the cost of sea water desalination has dropped significantly. This has prompted coastal communities in Florida, Texas, Southern California and elsewhere to look at the virtually unlimited supply of water at their doorsteps, the oceans. Distillation and a process known as reverse osmosis are the two most commonly applied methods for the desalination of sea water. They will be described below in the section on tertiary treatment of waste water.

About two percent of Earth's fresh water supply is contained in polar ice, but using it would require overcoming an economic hurdle. The cost of transporting and melting the ice does not appear to be practical at this time. Other strategies such as collecting water in regions where it is usually abundant and delivering it by pipes or canals to areas that are short of water have been used in some areas, such as California. However, economic and political forces have made that approach increasingly difficult. In other plans, tankers or huge rubber bladders have been suggested as the means for transporting the water over ocean routes to water-poor regions.

The recycling of water is as old as Earth itself. Nature has been recycling water for eons. The water that falls as precipitation was at one time part of the ocean, which through evaporation rose and formed clouds and eventually fell as rain or snow to return to the ocean via streams and rivers to begin the cycle again. Human civilizations also have long recycled water. Over the millennia, water from agricultural overflow and human waste was emptied into the nearest body of water, a lake or river. That same lake or river probably supplied water for others at some other location. Such practices generally created few problems as long as the populations were small and scattered and the body of water was large, making the process of self purification possible. Organic matter in the water was broken down by microorganisms and pathogens either were killed by solar radiation or were simply diluted out to the point where their numbers were essentially harmless. As populations concentrated into urban areas, demands for safe water supplies intensified. The great 19th century cholera epidemics of Asia, Europe and North America made people aware of the existence of water-borne diseases and the urgent need to find means of preventing them.

While the recycling of water is an effective answer to present day water shortages, it suffers from many potential difficulties. When one examines the many uses of water, one sees some potential conflicts. The lake or river water that receives large amounts of domestic and industrial waste, laden with harmful microorganisms and chemicals, is the same water that will be used for drinking, cooking and bathing at some other place and time. Before it can be reused in this way, the water must be treated to make it safe or "potable". **Potable** water is defined as water that is free of harmful organisms and chemicals and is fit to drink. Treating water to remove cloudiness and unpleasant taste goes back many thousands of years and involved filtering it through beds of sand. Modern treatment methods for producing safe drinking water were developed early in the 20th century and also

76

involve filtration along with a newer innovation, disinfection. By the second quarter of the 20th century, the incidence of water-borne diseases in the United States such as typhoid fever and cholera had dropped to near zero in those communities that adopted modern water treatment methods.

Modern Drinking Water Treatment

Drinking water piped to homes and businesses can originate from surface water or groundwater. **Surface water** is taken from rivers, lakes or reservoirs while **groundwater** is pumped out of the ground by wells. About 65 percent of water supplied by public water systems in the United States comes from surface water, the balance originating from ground water. Many communities rely on a combination of the two sources.

Typical steps for treating drinking water are shown in Figure 3.1. If the source of the water is surface water, such as from New York's Pepacton Reservoir, Lake Michigan or the Colorado River, the water probably contains a large amount of suspended solids (mostly soil particles and plant debris) and dissolved minerals. In addition, it is likely the water had been used previously by other humans to dispose of their industrial and domestic waste and will contain varying amounts of dissolved chemicals and microorganisms, some of which may be harmful. All in all the water will require extensive treatment to make it potable. Larger particles are allowed to settle out by storing the water in reservoirs for a short period of time. The remaining smaller particulate matter is then precipitated by adding flocculating agents such as aluminum sulfate to the water. The **floc** (the particles that have settled out) is separated from the water and the water is then passed through a series of filters containing gravel and sand to remove the smallest particles, including most of the microorganisms.

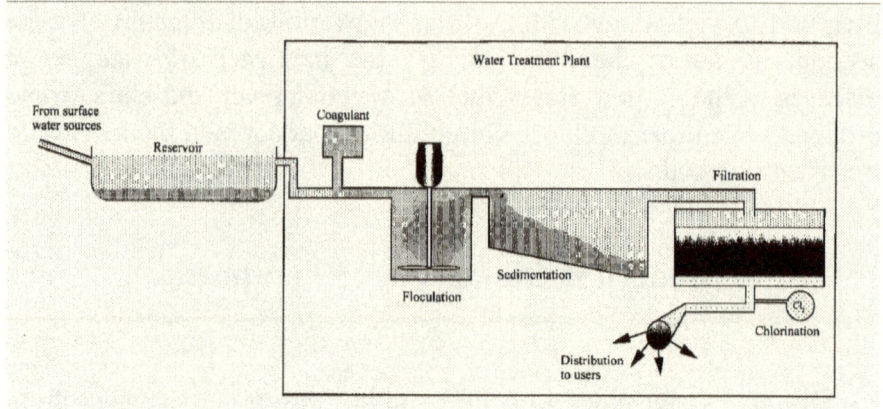

Figure 3.1. Diagram of a typical municipal water treatment plant.

Water is collected from various surface sources and stored in a reservoir to allow large particles to settle. The water is then delivered to a treatment plant where coagulants are added to aid in the removal of smaller sediment particles. The finest particles, including microorganisms, are removed by passing the water through a bed of sand. The water is finally chlorinated and delivered to users via water mains.

If toxic chemicals such as arsenic, chromium or nitrate are present in the water, it may be subjected to a biological treatment wherein microbial activity reduces the water's hazardous properties, or fed through a bed of ion exchange resins. Ion exchange resins are polymeric substances that have the ability to capture specific chemicals. To remove odors and off-flavors the water may be passed through a bed of coal or charcoal. Finally it is disinfected, tested and delivered to consumers via pipelines.

Groundwater that is pumped from deep underground sources is usually free of harmful microorganisms and except for greater amounts of dissolved minerals it is generally of better quality than surface water. Groundwater usually does not require all the treatment steps applied to surface water and is simply disinfected and piped directly to consumers. Groundwater originates from surface water that has percolated down through the soil until it reaches an impervious barrier, forming an **aquifer**, in a sense an underground lake. Under ideal conditions this natural filtration process removes most microorganisms and pollutants from the water.

The quality of groundwater in many areas of the U.S. has been deteriorating in recent times due to both natural and human causes. Fissures in the ground can allow contamination from leaking sewer pipes, septic tanks or treatment facilities to enter aquifers. Also, in agricultural areas where crops are heavily fertilized or there is an excessive accumulation of animal waste from so called "factory farms", a buildup of chemicals such as nitrates may percolate into the groundwater or find their way into nearby lakes or rivers. It has been estimated that there are about 450,000 factory farms in the United States, producing over a billion tons of waste per year.

The EPA has found that wells in many states are producing water with nitrate levels that exceed drinking water standards. Nitrates are especially harmful to babies. Nitrates reduce the ability of a baby's blood to carry oxygen, resulting in cyanosis (lack of oxygen in the blood) and possible brain damage. In certain areas of California, where almost overnight large areas of agricultural land have been converted to urban development, residents are particularly exposed to groundwater with high nitrate as well as pesticide concentrations. Groundwater may also contain other chemical pollutants that have seeped into the ground from leaking fuel or chemical storage tanks (the most common source of groundwater pollution), unlined waste dumps or from the remnants of missile manufacturing and testing. The use of microorganisms to remove chemical pollutants from groundwater will be covered in Chapter 5.

The development of methods for producing safe, clean drinking water is one of humankind's most significant achievements. But even in modern times when water has been treated to current standards, there have been numerous instances where treatment systems have failed and people have became ill from drinking the water. The most dramatic episode in recent years occurred in Milwaukee, Wisconsin in 1993. The intestinal protozoan parasite *Cryptosporidium* found its way into the city's water supply and apparently survived the usual purification steps. Over 400,000 people became ill, with 4,400 requiring hospitalization and over 50 deaths. On a national level, the Centers for Disease Control and Prevention have estimated that nearly

1,000,000 cases of various water-borne illnesses occur each year in the United States, resulting in at least 900 deaths.

Biofilms as Water Purifiers

What's a Biofilm?

No, *The Life of Emile Zola* is not an example of a biofilm. A biofilm is a special type of microbial community. We have mentioned on several occasions that most microorganisms are single-celled. That immediately raises the impression that an individual bacterium may spend its life suspended in the water of a lake or in the intestine of an animal essentially isolated from the other organisms around it, carrying out its independent existence like some microbial Flying Dutchman. While this may be true in some instances, it is believed that in nature most microorganisms form aggregates or clumps. The aggregates may consist of millions of cells of a single species or a mixture of two or more species. There are a number of advantages for a microorganism to become a member of an aggregate. An aggregate tends to trap nutrients, and the members are protected from predators and toxic chemicals. Most significantly, members of such an aggregate may improve their survivability through various symbiotic relationships that involve cross feeding, exchange of genetic information and communication.

Cross-feeding means a bacterial cell may overproduce a certain nutrient (an amino acid or a vitamin, for example) that leaks into the surrounding medium and is used by a second bacterium. In exchange the second organism may release a different nutrient that the first organism can use. The end result is that both species of bacteria can survive while neither might alone.

Microbial aggregates can be freely suspended in a liquid habitat such as lake or river water or more often, attached to a surface. When attached to a surface the accumulation of aggregates is referred to as a **biofilm**. Biofilms are frequently found on a wide variety of natural and artificial surfaces that are continually wet, such as boat hulls, the inner surfaces of pipes and cooling towers, food manufacturing equipment, submerged rocks, teeth, contact lenses, the lung tissues of cystic fibrosis patients and implanted medical devices like catheters and artificial joints. In each of these instances, individual bacteria initially attach themselves to the surface through various **adhesins**, molecules that have adhesive properties. Contact with the surface initiates rapid reproduction, resulting in the formation of a microcolony. Chemical signaling among members of the microcolony, known as **quorum sensing**, then triggers the establishment of a mature biofilm consisting of scattered colonies each containing millions of members (Figure 3.2).

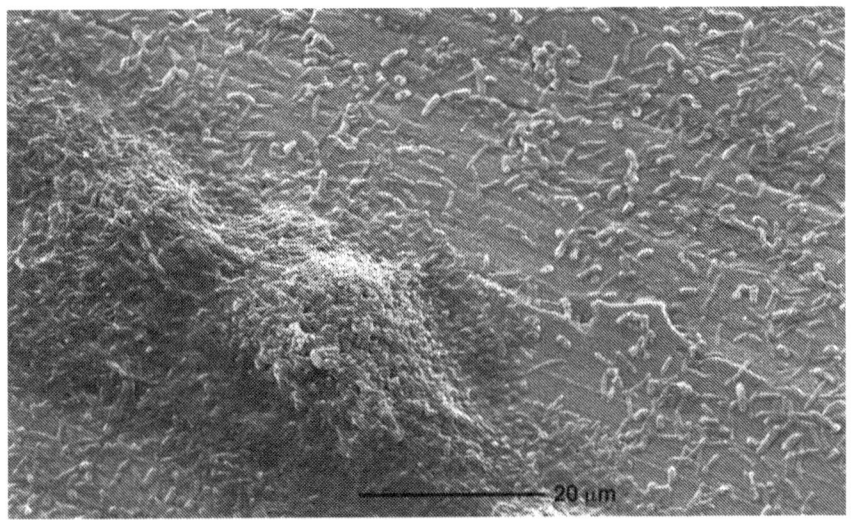

Figure 3.2. Electron micrograph of bacteria that had formed a biofilm
on a piece of stainless steel submerged in potable water for one week. Photo: Centers for Disease Control and Prevention/Janice Carr.

Members of the biofilm release large amounts of sticky polysaccharides which form a semi-rigid matrix that stabilizes the biofilm. Water channels between the colonies in the matrix function as conduits for nutrients and waste materials. In many natural settings, other bacteria, algae (if light is available), protozoa, worms, insects and other higher forms may join the bacteria that began the biofilm. A biofilm may become several millimeters thick, becoming what is referred to as a **microbial mat**. These mats sometimes can build up to nearly a meter or more thick to become **stromatolites** (Figure 3.3). Fossilized stromatolites have been found in several regions of the world and some appear to be over 3 billion years old, leading to the conclusion that life on Earth learned to socialize very early.

Figure 3.3. Stromatolites discovered on the sea floor off the Bahamas.
Photo: OAR/National Undersea Research Program (NURP)

In the case of a boat hull, barnacles, mussels and other marine organisms also may become part of the biofilm. Biofilms can frequently lead to serious problems. The buildup of organisms on the boat hull eventually adds weight and increases resistance to movement through water, impairing the operation of the boat and requiring periodic removal. A similar situation often occurs with water intakes for power plants and other industrial facilities. The accumulation of organisms in the pipe can significantly reduce its carrying capacity. To prevent the formation of biofilms, so-called anti-fouling paints can be applied to boat hulls, and strong disinfectants may be flushed periodically through water intakes to discourage microbes from starting a biofilm. The application of anti-fouling paints and disinfectants must be carefully monitored, for they usually contain toxic chemicals which when released into the surrounding water can adversely affect plants and animals in adjacent habitats.

Biofilms also are involved in certain aspects of human health. As mentioned above, biofilms can occur in the body on implanted medical devices such as artificial joints or catheters as well as natural surfaces like teeth. As much as 65 percent of human bacterial infections may involve biofilms. Biofilms afford pathogens extra protection against the body's normal defense mechanisms such as antibodies and phagocytosis, as well as against any introduced antimicrobial drugs, allowing the organisms to grow unhindered. The treatment of infections associated with biofilms is clearly a difficult medical challenge.

However, bofilms can be applied to solving a number of environmental problems. A bed of gravel or plastic pieces, for example, on which a biofilm of appropriate organisms has formed, can remove pollutants from water that flows through the bed. The microorganisms in the biofilm use the pollutants as sources of energy and material, and the polysaccharide matrix traps heavy metals. This principle is used in applications such as aquarium filters and in the treatment of waste water and urban runoff.

Waste Water Treatment

About 25 percent of households in the United States, mostly in rural regions but in some suburban areas as well, dispose of their sewage waste in septic tanks. A **septic tank** is an underground concrete or metal chamber where sewage waste is collected and allowed to decompose. Solids settle to the bottom of the tank while the liquid portion is drawn off into perforated pipes that spread out into an adjacent **leach field.** Liquids from the pipes filter into the ground and undergo natural purification. The solid material, which has accumulated in the tank and has undergone microbial decomposition, must be pumped out periodically for disposal. The safety and efficiency of septic tanks varies according to such things as climate and terrain. High water tables increase the likelihood of

David M. Carlberg, Ph.D.

contamination of water supplies and overly wet climates inhibit proper release of the liquid waste, causing backups and overflows. In addition, lateral movement of effluent may reach nearby flood control channels, rivers, beaches or other open bodies of water used for recreation. Many county and city governments are moving to phase out septic tanks and connect the households to waste water treatment systems, but costs have been the principal deterrent. In the meantime the EPA has launched a program to help communities teach their residents proper maintenance of septic tanks via a CD called "Wastewater Month" (see Further Explorations).

In most urban areas, waste water from homes and businesses is collected by sewer mains and directed to waste water treatment plants. Waste water typically is passed through one or more of three principal treatment steps, called primary, secondary and tertiary (Figure 3.4).

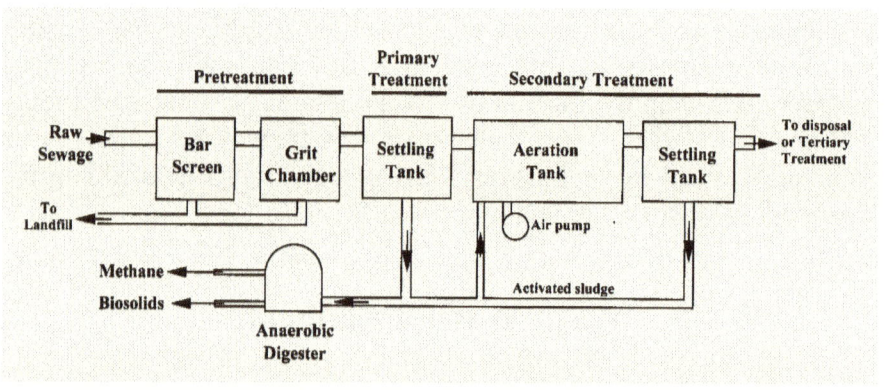

Figure 3.4. Layout of a typical waste water treatment plant.

Pretreatment

Waste water undergoes a pretreatment by passage through bars and screens to separate larger solids and grit (principally egg shell fragments, coffee grounds and sand) from the liquid portion. Material collected in these steps is mostly non-organic and not suitable for further treatment. It is usually transported to landfills.

Primary Treatment

Primary treatment consists of letting the raw waste water stand in large tanks to allow smaller organic solids either to settle to the bottom or rise to the surface where they are collected. Some facilities carry out **advanced primary treatment** in which chemicals called coagulants are added to the waste water to increase the amount of material that settles out. The solids that are collected undergo microbial decomposition in large anaerobic digesters, and then pass through a dewatering process where most of the water is removed. Methane gas is one of the byproducts of the anaerobic digestion step and can be collected and used as a fuel.

Biosolids

The partially digested and dewatered material that remains after anaerobic digestion is called Class B **biosolids** and can be used in limited agricultural applications, burned as a fuel, or simply buried in landfills. If the biosolids are further treated to reduce levels of heavy metals and microorganisms, it is designated as Class A and can be used in wider agricultural applications. In past years many seacoast cities discharged their biosolids into the ocean, a practice that was ordered stopped by the Environmental Protection Agency in the 1990s.

Eutrophication

The effluent from the primary treatment step still contains high levels of dissolved organic matter, small suspended solids and viable microorganisms. If that liquid is discharged into a relatively small body of water such as a lake, as the microorganisms in the waste water metabolize the organic matter, they will consume oxygen and soon the lake's oxygen level will be inadequate to sustain fish and other aquatic life. Dead fish begin to litter the shoreline and the lake is said to have undergone **eutrophication**, meaning excessive nutrients deposited in the lake have depleted the water of its oxygen and the lake has turned into an oxygenless desert (Figure 3.5). To avoid eutrophication the level of oxidizable organic matter in the waste water must be reduced before the water is discharged into the lake. The first step in determining if the effluent if safe to discharge into a body of water is to measure the waste water's biochemical oxygen demand, or BOD.

Figure 3.5. Tons of dead fish accumulate at the edges of a lake that has undergone eutrophication.
Photo: Visuals Unlimited/(c)Rob and Ann Simpson.

Biochemical Oxygen Demand (BOD)

The amount of oxygen needed by microorganisms to metabolize the dissolved organic matter in a sample of waste water is referred to as the **Biochemical Oxygen Demand** or **BOD.** The oxygen level in a bottle containing waste water is measured, then the sample is sealed and incubated at 20°C for five days. The bottle is opened and the oxygen level is measured once again. The difference in the two oxygen readings is called the BOD and can be used as an indication of how much oxygen would have been consumed by microorganisms if that waste water had been discharged into a body of water. If the BOD is considered too high, the waste water must be further treated to reduce the BOD before discharge.

Suspended Solids

The fine suspended matter in primary treated waste water is referred to as total suspended solids (**TSS**) and contains high levels of organic matter and microorganisms. The release of excessive TSS into a body of water adds turbidity to the water, reduces photosynthesis, interferes with respiration of invertebrates and fish, increases microbial loads and destroys the esthetic quality of the water. Approximately ninety percent of the microorganisms in the waste water remain attached to the suspended solids.

Secondary Treatment

The principal purpose of secondary waste water treatment is to reduce further BOD and TSS levels in the primary-treated water. During secondary treatment, microbial activity is encouraged by passing the liquid through a variety of devices such as trickling filters and aeration tanks. Trickling filters (Figure 3.6) consist of large tanks containing stones or pieces of plastic on which a biofilm has formed. Primary-treated waste water is allowed to flow down through the bed for maximum contact with the organisms in the biofilms. In aeration tanks, air or oxygen is pumped into the primary-treated liquid, creating a turbulent environment that supports rapid microbial growth. Most of the dissolved organic carbon is digested by the microorganisms and is released as CO_2.

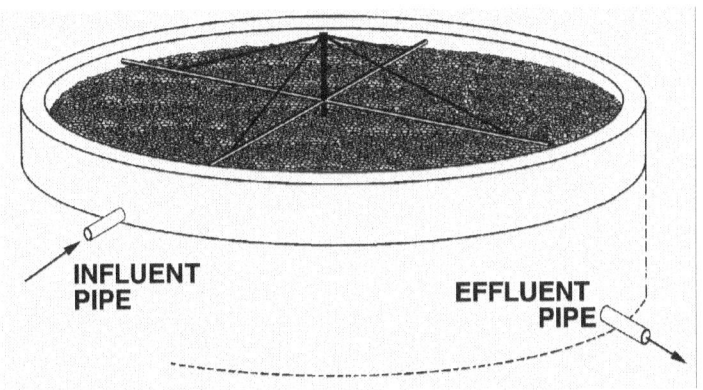

INFLUENT PIPE

EFFLUENT PIPE

Figure 3.6. Trickling filter.

Waste water from primary treatment step enters via the influent pipe, rises in center pipe and drops from four rotating distributor arms over a bed of rocks covered by biofilms of microorganisms. Treated water exits via effluent pipe.

Aggregates of viable microorganisms and undigested organic matter form in the aeration tanks of the secondary treatment step and eventually are allowed to settle out. The solids that are removed are called **activated sludge**, some of which is returned to the aeration tank to treat a new batch of waste water. The sludge produced in the secondary step is mixed with the solids from the primary step and transfered to the anaerobic digesters.

During secondary treatment, microorganisms consume much of the dissolved organic matter and the end result is that as much as ninety eight percent of the BOD has been eliminated. Most of the particulate matter is captured in secondary treatment as part of the sludge, bringing about a possible ninety percent reduction of the effluent's TSS. Because the BOD of the waste water now has been significantly lowered, the effluent is considerably less likely to cause eutrophication where it is discharged.

While reduction of suspended solids means most of the microorganisms have been removed, secondary treated water still contains high levels of microorganisms and is still unsuitable for uses where human contact is possible.

Tertiary Treatment

Waste water can be made safe for uses involving human contact and even potable by following the secondary step with tertiary treatment. Tertiary treatment can involve precipitation, distillation, filtration and disinfection. Chemicals are added to the secondary effluent to accelerate the removal of the remaining suspended solids, which either precipitate or are trapped by passage through fine filters by microfiltration. **Microfiltration** is a low pressure membrane filtration process that filters out very small particles including microorganisms, but not dissolved chemicals, meaning there is no significant BOD reduction. But by eliminating particulate matter, microfiltration greatly increases the efficiency of another widely used technique called reverse osmosis and often precedes it.

In **reverse osmosis** (Figure 3.7), water is separated from most its contaminants. The process is called *reverse* osmosis because it involves moving water under high pressure across a membrane to the side that contains pure water, leaving most dissolved salts, organic matter and particulate matter on the other side of the membrane. In ordinary osmosis, water moves from the side of the membrane with low solute concentration to the side of the membrane with the higher concentration of salts.

Distillation involves applying heat to the waste water until it evaporates, leaving salts, particles and microbes behind. The resulting vapor is then condensed and collected, producing essentially pure water. Because they are both extremely effective in removing dissolved chemicals from water, reverse osmosis and distillation also are frequently used for the desalination of sea water.

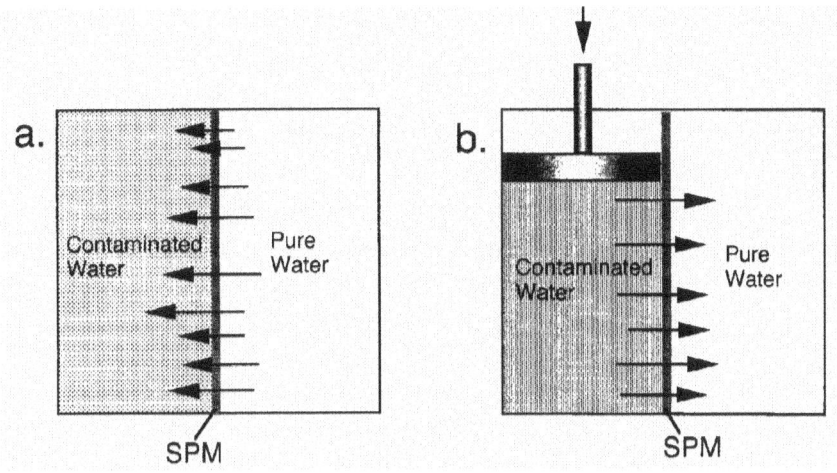

Figure 3.7. Comparison of normal osmosis (a.) with reverse osmosis (b.) In ordinary osmosis,
water flows to the side of the membrane with the highest concentration of salts. In reverse osmosis, water is forced under pressure from the salty side to the side containing pure water, leaving contaminants behind. SPM semi-permeable membrane.

Tertiary-treated water can be potable and safe for drinking and cooking. While tertiary-treated waste water actually can surpass all standards for drinking water, for psychological reasons as well as providing an extra degree of precaution, it generally has been used to replenish ground water supplies by percolation rather than being piped directly to consumers. Tertiary treated waste water is often more expensive than surface or ground water and therefore it has not been widely used for domestic consumption or landscape irrigation, but the realities of future water shortages are turning more communities to consider it.

Treatment Waivers

The Clean Water Act requires that all publicly owned waste water treatment facilities subject all of their waste water to full secondary treatment before releasing it into the environment. However, Section 301 (h) of the CWA provides that a coastal waste water treatment facility releasing effluent into the ocean is not required to subject its effluent to full secondary treatment if it can be demonstrated that due to the effects of dilution, ocean currents and the depth and distance from shore of its outfall and other criteria. . .

> ". . .such modified requirements will not interfere, alone or in combination with pollutants from other sources, with the attainment or maintenance of that water quality which assures protection of public water supplies and the protection and propagation of a balanced, indigenous population (BIP) of shellfish, fish and wildlife, and allows recreational activities, in and on the water."

The effluent must still conform to specific standards, including TSS, BOD and other criteria. Fewer than two dozen coastal sanitation facilities in the United States have been granted 301 (h) waivers.

Industrial Waste

Many industrial operations such as chromium plating, metal foundries and battery manufacturers produce large amounts of toxic waste in the form of heavy metals and acids. If allowed to flow into the sewers, the waste can interrupt the operation of a waste water treatment plant and flow into sensitive habitats. As you recall, a principal part of the treatment plant's workings, the secondary stage, depends on microbial activity. A large dose of chromium, mercury or other toxic chemicals can bring the secondary stage to a standstill. In addition, the toxic matter will eventually find its way to waters that receive the effluent, a river or ocean, causing severe impacts on local plants and animals. Such industries must find ways of removing the toxic materials from their waste water before the water is allowed to enter the sewage system. While certain industries may be obvious sources of toxic waste, some sources are not so clear. For example, recent studies have shown that dental amalgams (mixtures of silver and mercury) that are flushed down sewer drains from dental offices may be the most common source of mercury pollution in the environment.

Industries that produce waste water that is not toxic but high in BOD also must treat their effluent before it can be released into the sewer system. Cheese manufacturers, breweries, abattoirs and other food manufacturers characteristically produce huge amounts of high BOD-associated waste which often can overwhelm a treatment plant. Such industries must depend on oxidation ponds or other means to reduce BOD before allowing the waste water to enter the local sewage system. An oxidation pond is simply a large, shallow body of water in which waste water with a high BOD is allowed to undergo microbial digestion until the BOD has been lowered to an acceptable level.

Using Wetlands for Waste Water Treatment

Primary settling basins, secondary aeration tanks and other modern waste water treatment methods are nothing more than imitations of natural processes. Additional treatment of waste water can be achieved by passing it through a wetland. A wetland is an area of usually permanent, low-lying water containing characteristic soils and plants. Either natural wetlands or so-called **constructed wetlands** can be used to treat waste water following primary treatment. As the name implies, a constructed wetland is a human-made structure that is designed to simulate a natural wetland. One difference between a natural wetland and a constructed wetland is that the plants that are placed in the constructed wetland are chosen specifically for their ability to reduce pollution.

The vegetation of a wetland for water treatment is a critical factor in its success. Cattails, rushes and other common wetland plants play a number of roles. Once established, they take up metals and they act as barriers, slowing the flow of water and increasing the **retention time**, the time a given volume of water spends in the wetland. Lowering retention times aids the sedimentation of suspended solids and allows sufficient time for biological and chemical purification to occur. Vegetation also contributes directly in improving water quality by taking up nitrate and phosphate from the waste water. Microorganisms in the silt and attached to vegetation metabolize organic pollutants and immobilize heavy metals. Most pathogens are diluted out or die from exposure to sunlight. The water that leaves the wetland is considerably cleaner than the waste water that entered it.

Constructed wetlands intended to treat waste water can involve either **subsurface flow** or **surface flow.** The sole purpose of subsurface systems, also called root-zone systems or rock-reed filters, is to treat waste water. However, a number of cities and towns across the United States have discovered the additional advantages of utilizing surface flow wetland systems to treat their waste water. Not only is surface flow considerably more efficient in removing pollutants and less prone to clogging, there is also the opportunity of establishing much-needed wildlife habitat. Cities and towns such as Orlando, Florida, Vermontville, Minnesota and Arcata, California have constructed such facilities that support large numbers of birds and other wildlife while improving the quality of water before it is released into the environment (Figure 3.8).

Figure 3.8. Waste treatment wetlands at Arcata, California.

The recreational and economic value of treatment wetlands as wildlife habitats offers an added benefit. Bird watchers, wildlife photographers, artists and other visitors bring added business income to a community. Wetlands are much less costly than conventional waste water treatment plants to build and to operate, provided a community has sufficient level, inexpensive land available. Depending on the degree of treatment of the incoming waste water and the discharge requirements of the wetland effluent, a wetland for waste water treatment requires from about 1.5 to 5 acres of actual submerged area per 100,000 gallons of waste water per day.

Reclaiming Waste Water

In many intensely urbanized areas of the world the supply of drinking water is running out. For example, Southern California receives some of its water from the Colorado River, some from

Northern California, and some from local, underground sources. In spite of the semi-arid climate, many communities in Southern California may get three quarters or more of their water from underground sources. It is anticipated that due to population growth and competition for water from other regions, all sources will be inadequate in the coming decades. In past years, nearly every state in the continental United States and many foreign countries have experienced periodic and sometimes severe water shortages due to droughts, and the future will be no different.

One alternative solution to water shortages is to recycle waste water rather than discarding it into a river or ocean. Waste water is a combination of domestic and industrial sewage and is approximately ninety eight to ninety nine percent water and one to two percent suspended solids and dissolved organic and inorganic matter. While in the past that water has been treated and then summarily discarded into the nearest receiving body of water, that is changing. Los Angeles County treats about 500 million gallons of waste water per day of which over 180 million gallons per day are reclaimed. In Florida, waste water reclamation facilities are scattered throughout the state, resulting in the reclamation of nearly 600 million gallons of water per day. Most of the reclaimed water in these instances is not sufficiently pure for domestic use. However it can be used for landscape and agricultural irrigation, industrial and recreational applications and wildlife habitats, conserving high quality water for domestic use.

Through tertiary treatment, waste water can be made pure enough for drinking and cooking, but most commonly it is pumped into ground water supplies, a process known as **indirect reclamation**. (**Direct reclamation** would involve delivering the reclaimed water directly to consumers). For example, in 2001 Orange County, California, launched a $352 million Groundwater Replenishment Program in which tertiary treated waste water will be injected into the county's ground-water supply. The system will take secondary waste water and pass it through a series of tertiary steps involving microfiltration, reverse osmosis and ultraviolet disinfection. The project will ultimately produce 30 billion gallons of drinkable water per year.

The prospect of using reclaimed waste water directly for drinking and cooking has not been an easy sell to the general public. Even though that water would far exceed health standards, most consumers have not accepted the idea. Ironically, water that is presently delivered to homes in most communities around the country probably already contains recycled waste water from either indirect reclamation or from surface water that had been used in the past to dispose of waste water. In a recent survey conducted by the Institute of the Environment at the University of California at Los Angeles, no more than 18 percent of Los Angeles homeowners would find reclaimed waste water acceptable for drinking, and no more than 25 percent for cooking. However, most respondents (over 85 percent) would use it outdoors for washing cars and watering landscapes. Thus public recalcitrance appears to be the greatest deterrent to using more reclaimed waste water.

Water-Borne Diseases

Humankind was slow in recognizing that certain diseases could be transmitted by contaminated water, and that such risks could be avoided by relatively simple water treatment. The turning point appears during the cholera epidemic of 1849 that started in India and spread throughout Europe. In that year, John Snow, a English physician, noticed that in some sections of London the epidemic hit certain households harder than others. Snow discovered that those families that suffered the highest number of cholera deaths were getting their water from a particular pump on a popular thoroughfare known as Broad Street (now Broadwick Street). But employees in a brewery across the street from the pump showed few if any signs of cholera. They depended on a separate, much deeper well for their drinking water. Snow concluded that the water from the Broad Street pump must have been the source of the disease. By simply removing the handle from the pump, which prevented people from using it, Snow probably saved innumerable lives. Then in 1854 another wave of cholera hit London. This time Snow's attention was focused on a particular water company ("A") that pumped its water from the Thames River at a point near the center of town. Its customers were seeing about eight times more deaths from cholera than those of a competing water company ("B"), which also got its water from the Thames, but much further upstream.

Snow's investigations were conducted decades before the germ theory of disease was established. He had little inkling what the exact cause of the disease was, only that the excrement of cholera patients was highly infectious and that company A was getting its water from a point on the Thames River near where London disposed of its sewage. Similar to the situation with the Broad Street pump, Snow's only conclusion was that the 1854 cholera epidemic was caused by water that was contaminated with human sewage. By convincing families to avoid water from company A, Snow probably saved many more lives. As a result of this work, John Snow is frequently referred to as medical history's first epidemiologist.

Cholera is called a water-borne disease because it is primarily transmitted from person to person by contaminated water. An infected person releases billions of cholera bacteria in his/her excrement. Because of the practice of recycling water, the bacteria may find their way into the water supply of other unsuspecting individuals, continuing the cycle. Once the germ theory of disease was established in the late 1800s, there was a much clearer understanding that water supplies were possible sources of diseases such as cholera, typhoid fever and dysentery. The methods used by the ancients to rid their water of unpleasant tastes and turbidity were soon used to treat water to make it free of harmful microorganisms. The methods simply followed nature: allow the water to filter through layers of sand, trapping microorganisms in the process, followed by chemical disinfection (first introduced in the United States in 1908) to kill most of the microorganisms that may have escaped the filtration step or may enter the distribution system after the water was treated. See Chapter 2 for more discussion of water disinfection.

Water Testing

With the development of methods for treating drinking water early in the 20th century, a simple procedure was needed to test whether a water purification system was working and whether the resulting product was safe to use for cooking and drinking. Later in the century, concern also was focused on the health hazards of fecal contamination in recreational waters. Around 1905 a standardized method was proposed that detected a group of bacteria that were normally found in high numbers in human waste, so-called **coliforms**. Coliforms usually do not cause disease themselves; they are considered **indicator bacteria**, meaning their occurrence may suggest the presence of sewage contamination. Notice the word *may*. Coliforms are also found in other natural habitats, such as plants and soil, that are normally not contaminated with human sewage. As part of their normal intestinal contents wild animals also carry coliforms that can end up in water supplies. In spite of this obvious shortcoming, the **total coliform** (**TC**) assay has survived nearly 100 years as a measure of water quality. A somewhat more specific test detects the presence of **fecal coliforms** (**FC**). These are coliforms that can grow at temperatures higher than other coliforms and thus represent inhabitants of the intestines of warm-blooded animals, including humans. While the fecal coliform group of bacteria consists of several species, the model fecal coliform is the bacterium *Escherichia coli* (**EC**)*,* or *E. coli* for short. A third test that is recommended by the EPA depends on detecting the presence of another intestinal bacterium of warm blooded animals known as **enterococcus** (**ENT**). Testing for ENT is especially useful for ocean recreational water since these organisms appear to survive longer in salt water than other indicators, making the test more sensitive.

An *E. coli* to Avoid

While *E. coli* is normally considered a harmless intestinal inhabitant, certain strains of this bacterium have acquired virulence factors that have converted it into a potential killer, especially of young children. One particular strain, called O157:H7, has been implicated in a number of outbreaks of colitis (inflammation of the colon) resulting in some deaths. The organism apparently acquired its lethal proclivity through genetic exchange with a closely related bacterium that is responsible for bacterial dysentery, *Shigella*. Strain O157: H7 appears to be a common inhabitant of the intestines of beef cattle, which explains why many of the outbreaks occurred among people who had consumed under-cooked hamburgers. This strain has also been detected occasionally in drinking water, raw milk, unpasteurized apple cider and on fresh produce.

The testing for the presence of indicator bacteria involves collecting water samples and inoculating them onto special bacteriological media. The bacteria are then identified according to how they react to the various media and their microscopic appearance. For example, one of the major characteristics of coliforms that helps identify them is their ability to form gas when growing in a medium containing the sugar lactose. Gas formation is easily observed as a trapped bubble in liquid culture media (Figure 3.9a).

The mere presence of coliforms in water is not sufficient reason to raise concern, since as has been noted, coliform bacteria are commonly found in nature from non-human sources. Therefore *how many* coliforms are present in a water sample has been used to determine whether a potential hazard to human health exists. Measuring bacterial populations is a labor intensive activity, and since a typical public health agency may be collecting dozens or hundreds of samples a day, some short-cuts had to be developed. One such labor-saving device is the Multi-Tube procedure. Several tubes of broth medium containing the sugar lactose are inoculated with various volumes of a given water sample. For example, 10 tubes might be inoculated with 0.1 mL, another 10 with 1.0 mL, and another 10 with 10 mL. (mL = milliliter, about 1/28th of a fluid ounce) Following incubation the tubes are examined to determine how many are positive, that is, show the formation of gas (Figure 3.9a). The formation of carbon dioxide gas from lactose is one of the major characteristics that define coliforms and no other bacteria in a water sample that would form gas from lactose are likely to be present. By referring to a statistical table of **Most Probable Number (MPN)** values, one can then get an estimate of how many coliforms were in the water sample. The numbers are usually reported as the MPN of coliform organisms per 100 mL (MPN/100 mL).

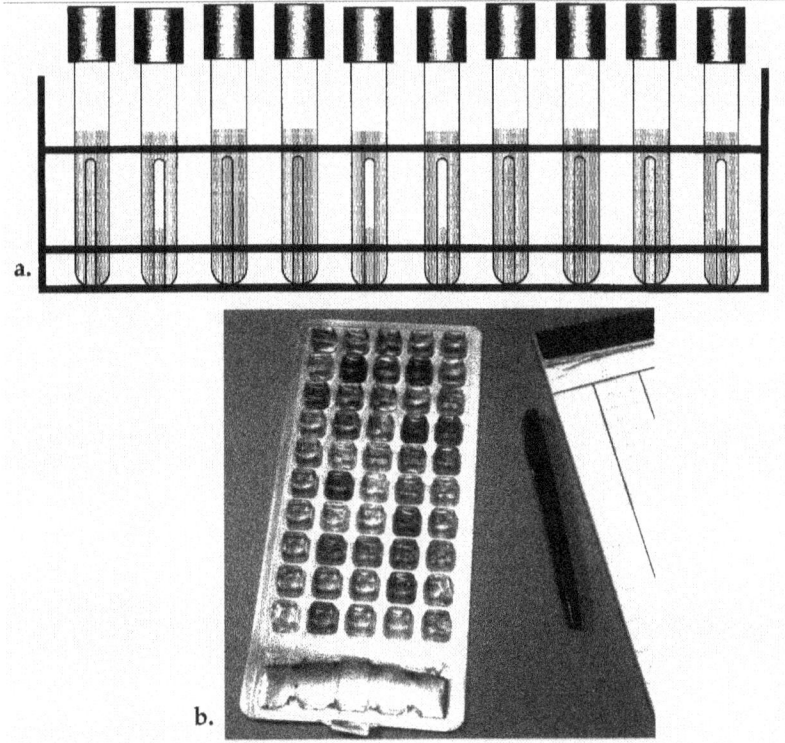

Figure 3.9. a. A multi-tube coliform assay.

Ten tubes of broth are inoculated with 0.1 mL of water to be tested. At the same time two additional sets of tubes are inoculated with 1 and 10 mL of the water sample (not shown). All tubes are incubated for 18 hours, after which each tube is examined for the presence of a gas bubble trapped in the small inverted tube. By referring to a statistical table, the number of tubes with bubbles in each set (4 out of 10 in this example) will give an estimate of the number of coliform bacteria (known as the "Most Probable Number") in the water sample per 100 milliliters. b. A Quanti-Tray coliform assay.

More recent methods based on statistics similar to those used in the multi-tube method have made water testing even less labor intensive. For example, a system called Quanti-Tray® (IDEXX Company) (Figure 3.9b) involves placing a 100 mL water sample into a tray containing 51 small wells. The sample is distributed among the wells, and the tray is sealed and incubated. The tray contains a differential medium that shows a color change if coliforms are present. and fluorescence under UV light if *E. coli* specifically is present. Another similar test detects enterococcus. By counting how many wells are positive and then referring to a computer-assisted statistical routine, an estimate of the number of indicator bacteria in the water sample can be made.

Another technique involves passing water samples through membrane filters that have porosities small enough to trap bacteria (Figure 3.10a). The filters are then placed on differential media that support the growth of the specific indicator bacteria and the resulting colonies that are characteristic of EC, TC or ENT are counted and related back to the volume of water sampled (Figure 3.10b). From these results the number of colony forming units (CFU) per 100 mL is calculated.

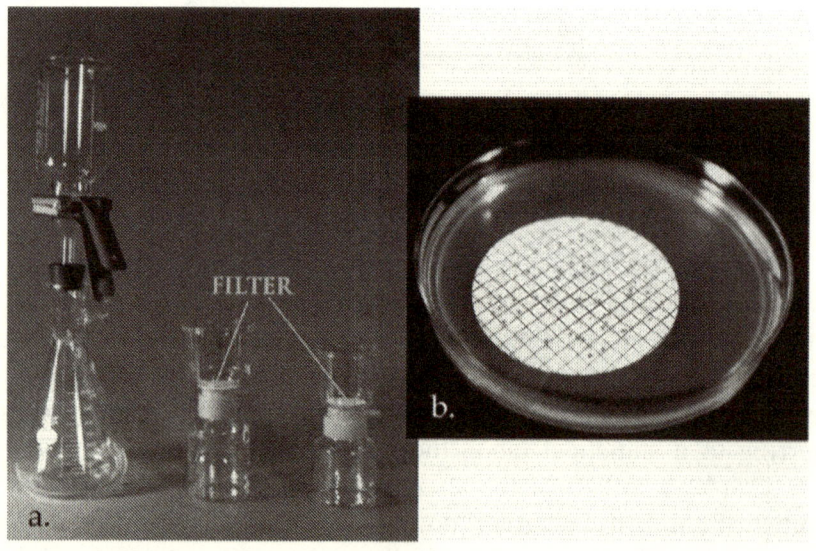

Figure 3.10. Examples of filter apparatuses used for testing water.

A carefully measured amount of water is passed through the filter, which traps individual bacterial cells. The filter is then placed onto a medium to allow formation of colonies (b), which can be counted to determine level of bacterial contamination.

In determining the safety of recreational waters, performing both FC and TC tests offers an additional source of information as to the possible origin of the bacterial contamination. As a general rule, if one calculates the ratio of FC to TC, anything over 0.1 suggests human origins. In addition, for those areas where ENT counts are made, some microbiologists claim that ratios of FC to ENT greater than 4.0 strongly suggest contamination of human origin, whereas ratios less than 0.7 denote non-human sources. These ratios are based on figures determined from various human and animal sources. Most wild and domestic animals show a FC:ENT ratio of less than 0.6 while the ratio in humans is about 4.4. The reliability of this method has not been fully tested. Unfortunately there is no uniformity among jurisdictions as to which of the various indicator bacteria is to be used to determine water quality. Some coastal states rely on FC or ENT counts or both for ocean samples, while some depend on EC counts, which is recommended by the EPA for fresh water. Many states have yet to adopt the EPA recommended bacteriological standards for recreational waters.

Rather than rely on the presence of coliforms to detect polluted water, why not test for specific human pathogens directly? There are several reasons why that is impractical at this time. Human pathogens, even in water polluted with waste, are usually in relatively small numbers compared to the enormous populations of the harmless intestinal bacteria, making it difficult to detect the true pathogens. Even if the pathogens were in larger numbers, there are no simple and inexpensive tests, such as those for coliforms, that are sensitive enough to detect water borne pathogens. To make matters worse, microbiological research has shown that when released into the ocean, pathogenic bacteria often lose their ability to grow on laboratory media, but they still retain their ability to cause disease. For example, a water sample may contain pathogenic cholera bacteria, but they might remain undetected if the sample were tested by standard culture methods for cholera. More advanced techniques are being developed that can pick out the pathogens in water samples. These are described below.

David M. Carlberg, Ph.D.

How Many Coliforms Is Too Many?

Since a satisfactory technique appropriate for detecting the specific presence of human waste in water is not presently available, we are limited to relying on the decades-old method of simply counting coliforms. That leaves the question, at what point is the water that was sampled deemed unsafe? That depends on what the water is being used for. If it is for drinking, the limit set by the EPA is zero per 100 mL. For swimming, shell fishing, boating and other uses, the limits may vary depending on the use and regulatory jurisdiction. For example, Table 3.2 lists the California ocean water standards for contact sports (swimming, water skiing, snorkeling). Notice that the standards allow for single samples to exceed means of extended samplings, giving local health agencies some flexibility in dealing with inevitable variability in sampling. The standards are based for the most part on epidemiological studies that matched frequencies of illnesses with coliform counts, or in the case of marine waters, with enterococcus counts. A tolerable maximum frequency of illnesses was then selected and the coliform or enterococcus counts that corresponded to that level of illnesses was set as the standard.

Table 3.1. Example of bacteriological standards for water contact sports.

Figures are for number of organisms per 100 mL of sample.

California Ocean Water-Contact Sports Standards

Single sample standards:	
Total Coliforms	10,000
Fecal coliforms	400
Enterococci	104
Fecal/total coliform ratio	1000 total coliforms if ratio exceeds 0.1
30 day geometric log mean standards of 5 weekly samples	
Total coliforms	1000
Fecal coliforms	200
Enterococci	35

Source: The California Code of Regulations - Title 17 and the California Health and Safety Code.

Once the results of water tests are known, any subsequent action to be followed usually depends on the circumstances. If drinking water exceeds the standard, users are notified to boil their water before using it, additional chlorination may be ordered, or both. And of course investigations are launched to locate the source of the pollution. In the case of water in a swimming pool, the facility might be closed and additional chlorination applied until test results drop below limits. If the offending water is at a swimming beach, the area might be posted with warning signs or the beach may be closed entirely to water-contact sports until bacteriological counts drop below limits.

In some areas, such as California, the response to high bacteriological counts at swimming beaches is specifically mandated by state law, and the authority rests with the local health agencies. If high counts can be directly attributed to sewage, such as the obvious failure of a nearby sewer main, or indirectly based on a microbiological signature that suggests human origin, the beach must be closed to all water-contact activities. If high counts do not appear to be associated specifically with human waste, the beach is merely posted with warnings informing the public that high bacterial counts were detected and participating in water-contact sports are at one's own risk (Figure 3.11).

Figure 3.11. Warning sign of high bacterial levels
Left: Warning sign of high bacterial levels; water remains open to swimming
Right: Sign indicating water closed to swimming due to sewage contamination

One major problem with routine bacteriological testing of water that has hounded public health agencies for decades is that there is a delay of up to 24 hours or more before the results of any tests are available. That is because the bacteriological tests currently used involve the growth of the organisms, and as you learned in Chapter 1, that takes time, at least 18 to 24 hours. Depending on how often sampling is conducted, the surf at a beach may have been contaminated for at least a day prior to the discovery of the high counts. The beach is posted with warning signs and sampling is continued. When counts fall below accepted levels, which will not be known for another 24 hours or more after subsequent samples were taken, signs can be removed. With the more advanced detection methods described above, such as DNA analysis, results may be obtained much quicker and postings can be done on the same day the sample was collected. But unfortunately these methods are not yet approved for water quality testing.

Studies of ocean bacteriological data from Southern California beaches suggested that testing water once a day may not be sufficient to get an accurate picture of the safety of the water. It was discovered that bacteriological counts may fluctuate widely during the course of a day. A sample taken at 8 A.M. may indicate the water as safe, one taken at 10 A. M. may show otherwise, and the noon sample may show the water as safe once again. Continuous monitoring using present methods would be enormously expensive if not impractical, since there still would be the 24 hour delay to deal with. University of California at Irvine Professor Stanley Grant, the principal investigator in this work, suggested perhaps semi-seriously that once instant testing is available, green and red lights similar to traffic signals may have to be installed at beaches to alert bathers of the minute to minute safety of the water.

Drinking water in urban areas is continually tested by water purveyors, and as a general precaution against delays in detecting contamination, water supplies should be routinely disinfected before they are piped to consumers. Thus if the water has been contaminated, the disinfection will eliminate it immediately, and the testing will trigger a search for the source of the contamination. Federal law requires all water utilities to supply written water quality reports regularly to all their customers. In rural areas where there is no central water supply and private wells are the source of water, it is the responsibility of consumers to have their water tested periodically to confirm its safety.

David M. Carlberg, Ph.D.

Source Tracking

Once evidence of possible fecal pollution is confirmed in recreational waters, the next step is determining its source so remediation can occur. However, none of the routine procedures described above can accurately differentiate human from non-human sources of pollution. (The hazards to human health of non-human fecal organisms are still being debated.) On first observation, all coliforms look alike regardless of source, and more advanced methods are necessary to distinguish sources of microbial pollution. Techniques are under study to determine the source of the coliforms or other bacteria by antibiotic resistance patterns, DNA analysis or other methods that can detect those organisms specifically associated with human waste. For instance, some experiments have shown that fecal pollution from human and farm animal sources generally show patterns of antibiotic resistance different from those of bacteria from wild animals. Also, bacteria typically associated with sewage are thought to have unique DNA sequence patterns that are different from DNA sequence patterns of bacteria in urban or rural runoff. Information of this type is often referred to as signatures or fingerprints. Called **Bacterial Source Tracking (BST)** or **Microbial Source Tracking (MST)**, these techniques show promise but are still in early stages of development.

DNA analysis is an example of an advanced technique for detecting microbial pollution. Briefly, a DNA analysis might be conducted as follows. A water sample is filtered to capture all the microorganisms in it. The cells may be cultured to increase their numbers or the cells may be concentrated directly by filtration or centrifugation. Then DNA is extracted from the sample. The DNA may be subjected to a technique called **Polymerase Chain Reaction (PCR)** in which enzymes amplify the traces of DNA of specific organisms that are sought, such as *E. coli* or the cholera bacterium. The DNA is then enzymatically broken down into manageable sized pieces that are separated in a gel by an electric field, transferred to a membrane and then exposed to a **probe.** A probe is a short sequence of DNA that is unique for the given species of bacterium that is being sought. Like looking for a lock that matches a key, if fragments of DNA from the water sample match the sequence in the probe, the probe will stick ("hybridize") to the sample DNA and hence to the membrane, indicating that DNA from the bacteia under investigation was in the water sample. Some techniques may avoid the gel step and detect the presence of specific DNA directly at the PCR step. Most of these techniques are carried out by automated equipment and in some instances results are available in less than 30 minutes, not including the time necessary to culture the water samples if implemented.

A number of viruses also inhabit the intestines of humans and most of these viruses, such as hepatitis A and the Norwalk-like viruses (Noroviruses), are pathogenic. The detection of these viruses could be used to indicate the presence of human fecal waste, but unfortunately testing for specific human viruses is rarely done on a routine basis. It is expensive, time consuming and can only be conducted in specially equipped laboratories that are generally only found in university research labs, some private laboratories or larger health agencies. On the other hand, the detection of enteric-specific bacteriophages in water is relatively simple. As you recall, bacteriophages are viruses that infect bacteria. The occurrence of certain phages, called male-specific or F+ phages, seems to correlate with the presence of human strains of *E. coli.* (The reference to "male" refers to the bacterial mating type that the bacteriophages infect, not to the human host.) These phages could act as indicators of human fecal pollution. This approach is still under study.

Urban Runoff

Untreated sewage or effluent from waste treatment plants are not the only health hazards that may flow into recreational waters. Over recent years we have begun to realize that another major source of pollution is plaguing our water environments, urban runoff. Many believe urban runoff is the major cause of polluted recreational water. Urban runoff results when water from rain, snow melt, irrigation or excessive home or commercial use collects on impervious surfaces like driveways, streets and parking lots and flows into storm drains and flood control channels. From there the runoff eventually enters a nearby lake, river, estuary or ocean (Figure 3.12). There would be no problem if urban runoff were nothing more than pure rainwater (OK, we know that is a myth, but bear with me). Urban runoff can be laced with everything that ends up on the ground: fertilizers, pesticides, herbicides, detergents, tire rubber, brake-pad particulates, brake and radiator fluids, gasoline, motor oil, grass clippings, cigarette butts and other trash, soil particles, animal feces and animal carcasses.

Figure 3.12. A urban flood control outfall flowing into a coastal wetland.

The types of pollutants that appear in urban runoff generally can be divided into three major categories: chemical, biological and a third catchall category of general trash. The chemical portion of urban runoff can include sulfuric acid and toxic metals such as mercury, copper, cadmium, lead and zinc. Sulfuric acid results when coal or other fuels rich in sulfur are burned. Once in the atmosphere, the sulfur combustion products react with water vapor to form sulfuric acid, which in time may fall to Earth as precipitation (referred to as

acid rain). Toxic metals also are released from the burning of coal and from the manufacture of various metal products. The metals are released into the air through processes like welding and grinding and eventually fall to the ground by gravity or are washed out of the air by precipitation. Pollutants also find their way into the environment whenever products or items containing them are indiscriminately discarded. Toxic metals are used in many paints, pesticides and other household chemicals, as well as television receivers, computer circuit boards and monitors and leaded gasoline. Mercury is commonly found in electrical equipment, certain types of laboratory instruments, thermometers and fluorescent lamps. In addition, urban runoff can contain dozens of organic chemical pollutants that are found in pesticides, herbicides, industrial solvents, motor oil and gasoline.

The direct impact chemical pollutants have on the environment is primarily on the plants and animals in the ecosystem that receives the runoff. While the pollutants may be in sub-toxic concentrations when they reach a lake or estuary, they may accumulate in plant and animal tissue and are then passed up the food chain until they reach significant concentrations. This process is known as **biological magnification** or **biomagnification**.

One of the most infamous examples of biological magnification involves the pesticide DDT. Once DDT is dispersed into the environment, it enters the food chain and tends to deposit in the fatty tissue of animals. When the animals are consumed by predators at the next higher level of the food chain, the DDT is further biomagnified until it may reach levels that are 100,000 times greater than its original environmental concentration. One of the effects of DDT on wildlife is the disruption of normal egg development in birds. The presence of DDT in the environment almost led to the extinction of the Bald Eagle, the Brown Pelican, the Peregrine Falcon and other bird species in the continental United States. The use of DDT in the United States was banned in 1973 but its manufacture for export continued until the 1980s. There are still submerged deposits of DDT off the Southern California coast that resulted from years of it being dumped into the sewer by a manufacturing plant in Los Angeles. The DDT, insoluble in water, passed undegraded through a waste treatment plant and settled on the bottom of the ocean just off Los Angeles Harbor. While the DDT is slowly migrating northward into

Santa Monica Bay, its original quantity, over 260 tons, apparently has not changed appreciably since the mid 1980s, indicating an insignificant rate of physical, chemical and biological degradation. A plan is underway to place a cap of silt over the DDT to reduce its release into surrounding waters, a project that is funded by the manufacturer through a landmark legal settlement. In a preliminary test conducted in 2002, a small portion of the DDT was successfully covered with a layer of 1 to 1.5 feet of silt. These encouraging results will no doubt lead to more extensive remediation.

A similar scenario is being played out in New York. An estimated 50 tons of PCBs has accumulated over nearly four thousand acres of Hudson River bottom during a thirty year period. This toxic pollutant originated from a nearby manufacturing plant. In this instance the EPA has proposed to dredge the PCBs from the river bottom and ship the material to a hazardous waste site. See Chapter 5 for additional information about PCBs.

Humans may be exposed to hazardous levels of chemical pollutants such as DDT and PCBs if they consume fish or shellfish that contain magnified amounts of these pollutants. Such pollutants cause various physiological effects on humans and other animals. Some of these chemicals are known carcinogens, others are thought to impair the immune system or disrupt neurological or endocrine functioning. The latter condition may not appear for one or more generations following exposure. Because of high levels of such pollutants in certain local fishing grounds, officials have had to post warning signs informing the public of the dangers of eating fish caught in those waters (Figure 3.13).

**Figure 3.13. Sign at a popular Southern California fishing
pier warning anglers of contaminated fish.**

Chemical pollutants in urban runoff also can include nutrients that come from excessive fertilizer use or concentrations of animal waste from feed lots, dairies and factory farms. High in nitrogen and phosphorus, these materials can create serious consequences when released into natural habitats, leading to algal blooms and eventual eutrophication and massive fish kills.

Silt has been a major problem associated with runoff. Silt is fine soil particles, and it often occurs in runoff wherever the soil has been disturbed, such as from logging, cultivation, construction or unpaved roadways. Its impact on natural habitats is considerable. Besides destroying the esthetic quality of water by creating turbidity, excessive silt can cover the bottom of a river, wetland or lake, burying and destroying whole communities of plants and invertebrates, which ultimately disrupts fish feeding. In addition, the silt covers gravel beds that are critical for spawning of fish species such as salmon.

The most immediate effect of the biological portion of urban runoff is on human health. Urban runoff contains enormous numbers of bacteria, viruses and other microorganisms, some of which are human pathogens. How do human pathogens get into urban runoff? That is unknown, but there are several likely explanations. Many flood control channels are unlined, making it possible for seepage from septic tanks or leaking sewer pipes to enter the channels. In some communities the homeless use flood control channels to dispose of their waste, and to avoid sewer hookup fees, property owners have been known to discharge their sewage directly into nearby flood control channels.

Microorganisms in urban runoff also affect wildlife. For example, parasite-infected sea otters along the California coast are thought to have picked up the infections from urban runoff that contains cat feces, a known source of the parasite.

While typical urban runoff has long been shown to contain human enteric (intestinal) bacteria and viruses, it also may contain non-enteric pathogens like *Staphylococcus* and *Pseudomonas* that are more often associated with infections of the skin, eyes, ears and respiratory tract. The origin of these organisms is unknown, but their discovery was probably to be expected considering the types of infections that have been reported by those who swim near flood control channel outlets. An important study at Santa Monica Bay, California reported in 1999 involving over 10,000 beach-goers showed the nearer to flood control channel outlets that people swam, the more likely they would suffer various infections, often involving the skin, eyes, ears and respiratory tract as well as the intestine.

Trash from urban runoff also has a significant effect on the environment. In addition to its negative esthetic impact, small bits of non-degradable matter such as cigarette butts and fragments of plastic foam or plastic bags can be mistaken by animals as food, to cause serious if not fatal blockage of the digestive tract. For example, to sea turtles plastic bags look like jelly fish, one of their main foods. It is frighteningly obvious what the consequences would be if a turtle tried to swallow a plastic sandwich bag. One biologist reported observing an albatross with a green plastic toothbrush stuck in its throat. It had apparently mistaken the toothbrush for a fish but was unable to regurgitate it when the bird attempted to feed its young. The fate of the bird was not reported. In addition, animals are frequently entangled, sometimes fatally, in larger items that wash down with urban runoff such as discarded fishing lines or the plastic webbing that holds six-packs of soda cans or water bottles.

Treating Urban Runoff

Like waste water, urban runoff also can be subjected to various physical and biological treatments that reduce levels of pollutants. There is a number of approaches, generally known as **GMPs** (Good Management Practices) that can be applied to treating runoff. There are structural and non-structural GMPs. A structural GMP can be as simple as a row of sandbags around a flood catch basin to prevent silt and trash from a construction site from entering a flood control channel. More elaborate structural GMPs may be various types of traps and filters that can reduce the amount of silt that flows into streams, lakes and the ocean. Other devices capture trash or grease and oil. So-called biofilters are designed to reduce nutrients and organic pollution before they enter sensitive habitats. Excessive runoff can be avoided through efficient irrigation technology. For example, electronic units receive weather and other data and automatically determine optimum watering schedules for programmed irrigation systems.

Non-structural GMPs can involve educational programs to assist developers and home owners in reducing runoff through better project design or water conservation efforts. Various forms of visual reminders concerning dumping of trash into storm drains are another use of non-structural GMPs (Figure 3.14).

David M. Carlberg, Ph.D.

**Figure 3.14. Various forms of reminders not to discard
hazardous materials into storm drains.**
a. Markings on curb above catch basin. b. Refrigerator magnets.

Without disinfection, structural GMPs have little or no impact on microbiological contamination in storm water except for what is trapped on particulate matter. Many of these devices have other shortcomings. Their ability to treat large volumes of water is limited, meaning they are only effective during relatively dry weather, that is, during times when runoff originates from sources other than major rainstorms. Known as **nuisance flow**, the source of dry weather runoff includes overflow from irrigating landscape, washing cars, driveways and sidewalks, draining swimming pools and similar activities. The runoff from the initial one half to three quarters of an inch of rain from a major storm is known as the **first flush**. The first flush washes large amounts of trash and other pollutants that had accumulated since the last rain into the storm drain system. If filtration devices have been installed, they catch most of the pollutants from the first flush, but as rain flow increases, the GMPs usually cannot handle the greater volume of water. These devices have overflow features so that the excess water by-passes the filtration section.

GMP devices require periodic maintenance, involving extra costs to the responsible agencies. Although the initial cost and regular maintenance costs have inhibited the wide use of these devices, greater enforcement of federal regulations are compelling cities and counties to require them on all new construction or significant remodeling projects, or apply other means of pollutant removal such as wetlands.

The city of Santa Monica, California has constructed a facility to capture and treat urban runoff that normally would have emptied onto beaches of Santa Monica Bay, which host 50-60 million visitors a year. Known as the Santa Monica Urban Runoff Recycling Facility (SMURRF) (Figure 3.15), the operation can handle up to 500,000 gallons of runoff per day. Dry weather runoff is collected from storm drains and subjected to steps that remove grit and sand, oil and grease and fine suspended solids. The water is then disinfected by ultraviolet light and distributed to users for landscape irrigation and for toilets in dual-plumbed buildings.

**Figure 3.15. Santa Monica Urban Runoff Recycling
Facility ("SMURRF"),**
urban runoff purification facility in Santa Monica, California. Runoff
enters facility at the far end, passes through several filtration units to
remove silt and oil, and is then subjected to ultraviolet radiation in
unit closest to camera.

To protect recreational waters from the excessive bacterial
contamination in urban runoff, other Southern California coastal
communities in San Diego and Orange counties have begun installing
facilities similar to Santa Monica's SMURRF, but with a difference.
Rather than diverting the effluent for landscaping irrigation, the
effluents from these units are returned to their natural water courses
that eventually reach the ocean. While the ozone or ultraviolet
radiation features of these facilities are effective in reducing microbial
levels in the runoff, their operation raises some critical questions. As
we have pointed out on numerous occasions on these pages,
microorganisms play pivotal roles in food chains. How many
beneficial microbes that are part of these food chains are being

eliminated in these units? Are there ecosystems down stream that depend on a healthy, viable population of natural microorganisms for their well being? While these facilities certainly make swimming in the famous Southern California surf considerably safer, no biologist has been able to answer the question, are we trying to make our urban runoff too clean? In response to these concerns, some of these facilities are designed to allow a certain fraction of the runoff to by-pass treatment, thus permitting some microorganisms to reach the down stream habitats.

Where the space is available, constructed wetlands, grass fields and other natural areas also can be used to treat urban runoff. Constructed wetlands were described earlier in connection with the treatment of waste water. Because of considerably lower pollution levels in urban runoff compared with waste water, the land area sufficient to provide treatment of urban runoff is generally considerably less than what would be needed for treating comparable volumes of waste water.

The Irvine Ranch Water District in Orange County, California has proposed to build as many as 37 constructed wetlands to treat runoff from its 118 square mile watershed. Runoff from that watershed eventually empties into recreational waters of the nearby Pacific Ocean, waters that are enjoyed by millions of people annually. Water district officials estimate that using constructed wetlands to treat urban runoff is far less costly than traditional applying GMPs or diverting runoff into sewage treatment facilities as some communities are doing.

Diverting dry weather runoff through normal sewage treatment facilities is a relatively simple means of preventing pollution of nearby bodies of water. However, it is generally considered a temporary fix until other means of dealing with urban runoff are in place. But as with the GMPs described above, the practice also prevents potentially beneficial microorganisms and nutrients from reaching coastal ecosystems.

In order to divert runoff to waste treatment facilities, pumps must be installed in major flood control channels to deliver the runoff to the nearest sewer main. Dry weather volumes are usually a small fraction of the flow that such treatment facilities normally handle. At the first sign of significant rain, the runoff usually must be redirected to its normal flood control channels. Sewage treatment facilities cannot handle the volume of runoff from a major rainstorm.

Combining sewage and flood water is not a new idea. Many communities in the United States have long operated **combined waste water systems** in which runoff from streets and yards is permanently directed year round into the sewer mains. However, during major rainstorms when the volume of rain water exceeds the capacity of waste treatment facilities, the facilities are overwhelmed and rainwater, raw sewage and industrial waste are flushed into the body of water that would normally be receiving treated sewage. Such events are referred to as **Combined Sewer Overflows** (CSOs). Communities with combined systems are now under mandate from the EPA to minimize the impact of CSOs on the environment while long-term solutions are being implemented.

Testing Water from Urban Runoff

Since non-enteric human pathogens occur in urban runoff, a number of microbiologists have questioned the value of using water testing procedures that depend on the presence of enteric (intestinal) indicator bacteria such as EC, TC and ENT for testing recreational waters that receive urban runoff. While results of these tests correlate well among themselves, research has shown that there is little or no correlation between enteric bacteria counts and the presence of other human pathogens in recreational waters. Water samples may show low coliform counts but have high levels of skin or respiratory pathogens. Complicating matters is the fact that wild birds and mammals, which are frequently found around flood control channels, are known to carry the common enteric indicator bacteria. Thus high coliform counts may be due to the presence of wild birds, for example, in the absence of any human waste. Whether wild bird waste in recreational waters poses a threat to humans is still under study.

As discussed earlier in the section on waste water, the newer tests based on molecular techniques such as genetic probes and DNA sequencing have revolutionized the detection of microorganisms in the environment, but the procedures are expensive and generally limited to use by larger health agencies. Most importantly these tests have not yet gained official recognition as standard procedures. As more is learned about the identity of human pathogens in the environment, our confidence in water testing may improve, but microbiological water testing is still an inexact science.

Dead Zones and Wetlands

In several parts of the world, large expanses of coastal waters have experienced catastrophic losses of sealife. These incidences appear to be caused by excessive amounts of nutrients washing off adjacent land and accumulating in pockets offshore. The most extensive example of these so-called **dead zones** has been observed for at least 20 years off the coast of Louisiana, where each summer a large expanse of coastal waters becomes almost completely devoid of sea life due to low oxygen concentrations. In 2002, 22,000 square kilometers were affected, the largest area ever reported. It appears that the main cause is agricultural runoff rich in nitrates and other nutrients that originate from the farmlands along the Mississippi River. Because of the lack of strong currents in the Louisiana coastal portion of the Gulf of Mexico near where the river empties, nutrients build up, leading to eutrophication and the suffocation of essentially all forms of life. Boats that normally fished the affected area now have to travel considerable distances to avoid the dead zone.

Several plans have been proposed to solve the dead zone problem. One approach is to convince farmers in the Mississippi Basin (which drains 41 percent of the continental United States) to use less fertilizer, not a simple task. Another plan is to encourage farmers (through reimbursements) to restore wetlands adjacent to their fields to treat runoff before it reaches the river. Just as we saw in the treatment of waste water by passage through wetlands, excess fertilizer in agricultural runoff can also be consumed by the wetlands plants and microorganisms, significantly reducing nitrates and other nutrients. In addition to improving water quality offshore, the benefits to wildlife of extended bands of wetlands along the banks of the Mississippi are obvious. It is hoped that by 2015 the Gulf of Mexico dead zone will have been reduced by at least 75 percent.

David M. Carlberg, Ph.D.

Dangerous Algae

Algae are usually considered harmless organisms that grow on the sides of aquariums, fish ponds and swimming pools. There are, however, several species of marine algae that under certain conditions produce toxins that can be extremely debilitating or at worst, lethal if ingested. In most cases the disease syndromes are caused by consuming seafood that contain the algae. Occasionally our coastal waters experience huge increases in microalgae populations that are called algal blooms. **Red Tides** and **Brown Tides** are terms that are often applied to rapid population increases of algae that are the type that produce toxins. Red and brown refer to the coloration that the enormous numbers of algae impart to the water. The algae are taken up by fish or shellfish, which in turn may be consumed by humans as well as birds and marine mammals, all of which may be seriously affected by the toxins. In the spring of 2002 dozens of dead and sick dolphins and sea lions washed up on Southern California beaches, presumably having consumed fish, principally anchovies and sardines, that contained the toxic algae. The major cause of these blooms appears to be excess nutrients that have washed off adjacent land. Certain members of the algae-like cyanobacteria also can cause similar intoxications.

Algal and cyanobacterial intoxications can cause severe intestinal distress, including nausea, cramps and diarrhea, while some cases are accompanied by neurological symptoms including a burning or tingling sensation around the mouth and lips, headache, seizures, memory loss, respiratory failure and death. Four examples of bacterial and algal intoxications that have occurred in coastal waters of the United States affecting humans and marine mammals are: **Cyanobacterial Poisoning**, caused by eating fish that had consumed certain toxic strains of ***Cyanobacterium*, Ciguatera Fish Poisoning**, caused by consuming tropical reef fish that have fed on toxic algae, and **Amnesic Shellfish Poisoning** and **Paralytic Shellfish Poisoning**, both caused by consuming shellfish containing toxic algae. There are no antidotes for these intoxications, the only prevention is avoiding seafood that contains the toxin. Quarantines on harvesting shellfish during those times of the year when toxic algae are common have been effective in preventing widespread poisonings (Figure 3.16).

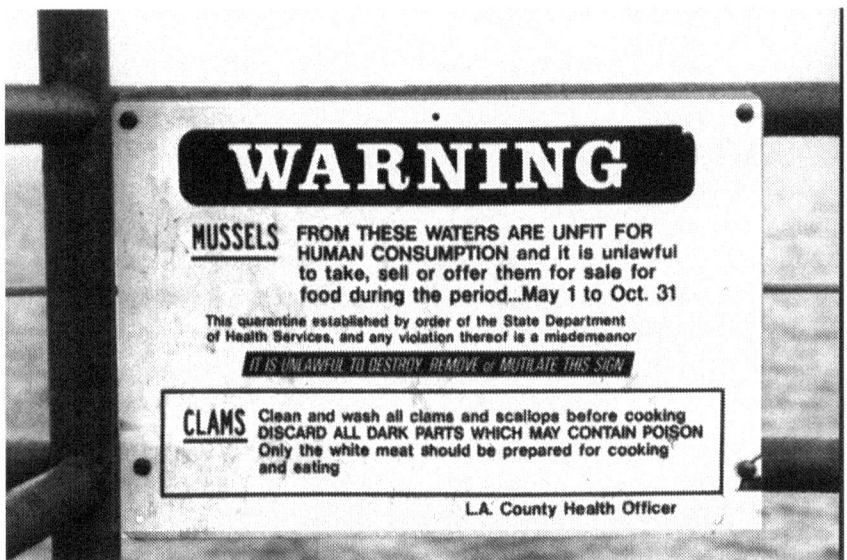

Figure 3.16. Warning sign posted during months when shellfish may contain dangerous toxins due to presence of toxic algae.

Starting in the late 1980s an alga known as *Pfiesteria* has been implicated in the deaths of over a billion fish in the bays and estuaries of the eastern seaboard of the United States The on-going story of *Pfiesteria* is an interesting example of how some scientific debates create considerable heat. At one point this one actually reached the floor of Congress. In a few words, scientists cannot agree on the specific cause of the fish deaths. The most likely answer is a toxin produced by the *Pfiesteria* organism, but some biologists have vigorously questioned that possibility, since no toxin has actually been isolated and identified. When water containing *Pfiesteria* is placed into an aquarium, all the fish present soon die. But are the deaths caused by the *Pfiesteria,* other organisms in the introduced water, or a combination of these? The complete answer to this fascinating mystery is far from clear. There have also been reports of human illnesses on exposure to *Pfiesteria,* but these have not been fully confirmed.

An Unexpected Pollution Source

A practice that has been around since humans first used boats for transportation may be another, unexpected source of water-borne pathogens. Most raw materials and manufactured goods are transported worldwide by ship. Cargo ships are most stable, that is, immune to capsizing, when fully loaded. As their cargo is removed at various ports along their routes, the lost weight is made-up by adding water to the ships' ballast tanks. Then, when the ships are reloaded with cargo, the ballast water is simply pumped out into the surrounding waters. It has been estimated that cargo ships release about 80 million metric tons of ballast water from overseas ports into waters of the United States each year. In 2001 scientists discovered that ballast water can contain enormous numbers of microorganisms, including cholera bacteria. It is thought that the cholera epidemic that has affected Peru and other regions of Latin America since the early 1990s arrived in the ballast water of a cargo ship from Asia and soon contaminated shellfish along the Peruvian coast. It has long been known that ballast water has introduced numerous invasive, non-native aquatic plants and animals such as zebra mussels and Asian shoe crabs into United States coastal and Great Lake ecosystems, but this is the first time there has been evidence that coastal waters may be invaded by pathogenic microorganisms from ship ballast waters.

A number of preventative measures have been proposed for reducing the introduction of live organisms into United States waters from ships' ballast tanks. One measure involves ballast exchange where ballast water is dumped at sea and replaced with mid-ocean water, but proof of compliance with this measure is almost impossible to acquire. A direct approach would be to disinfect ballast water prior to its release. However, ship operators have balked at the added cost and claim the decomposition of dead organisms in ballast tanks could result in creating anaerobic conditions, accelerating the tanks' corrosion and eventual failure (see Microbial Degradation of Concrete and Iron in Chapter 5).

Aquaculture

Aquaculture or **fish farming** originates from an ancient desire to have a convenient and reliable source of seafood by raising fish or shellfish in captivity until they reach edible size. Modern aquaculture is practiced world-wide, involves over a hundred species of aquatic animals and operates under a variety of commercial or research objectives. The kinds of fish raised may be for decorative purposes such as koi or tropical fish, or for food, like salmon and catfish. Species such as shrimp, mussels, abalone and oysters also are produced by aquaculture operations. Aquaculture generally involves maintaining adult brood stock to produce hatchlings. The young animals are kept in tanks, ponds or net pens and fed a healthful diet until they become large enough for their particular purpose. If fish are released into the wild to improve natural populations, the operation is often referred to as **fish ranching**.

Most of the catfish and trout consumed in the United States and about half of the shrimp and salmon come from fish farms. Worldwide production from fish and shellfish farms, which supply one third of the world's seafood, grew at the rate of ten percent per year during the decade of the 1990s. The proportion of seafood supplied by fish farms is likely to increase in the near future since the harvesting of many wild fish and shellfish species has leveled off, due for the most part to diminishing native populations. In some parts of the United States, the catching of certain species of fish has been banned entirely due to depletion. Aquaculture activities in the United States represent annual sales of about a billion dollars, but that is less than two percent of the world's total aquaculture production. Most United States aquaculture involves fresh water species such as catfish, trout and crayfish. Except for some salmon and shellfish farms, salt water aquaculture (**mariculture**) is relatively rare in the United States due in part to scarcity of appropriate coastal sites to establish fish farms. Such sites require unpolluted water with good circulation, proximity to transportation and protection from storm damage. Competition with other ocean oriented activities such as recreation, housing, shipping and wildlife protection has limited United States mariculture compared with the rest of the world. As a result, the United States imports a great deal of farm-raised seafood, which has resulted in a seafood trade deficit of over $6 billion as of 2002.

No such siting restrictions occur in other parts of the world, where mariculture operations have proliferated, supplying the world with over $50 billion worth of fish and shellfish (See Figure 3.17). However, coastal fish farms can create serious environmental problems. Because of the relative rarity of United States mariculture operations, the environmental impact caused by them has tended to be localized and relatively minor. Due to their greater numbers, the environmental problems caused by foreign mariculture operations have been considerable. Any detailed discussion of many of these issues would be beyond the scope of this book, but let us briefly mention a few examples:

•Escapees from fish farms may interbreed with native species, possibly harming native gene pools;

•Establishing a fish farm often results in the destruction of a natural coastal habitat and the native flora and fauna associated with it;

•Food for farm-raised carnivorous species like salmon usually contain fish meal and fish oil derived from forage species like anchovies and sardines. Because of overfishing in some areas of the world, these species are on the brink of extinction. Making matters worse, for certain farmed species, it takes 2 to 4 pounds of fish meal to produce one pound of farm-raised fish.

Three major areas, however, do concern us. 1) Fish farms generate large amounts of organic pollution; 2) fish farms can become breeding grounds of pathogens that may spread to wild fish populations; 3) animal or human waste is sometimes included in the fish food, opening the possibility of transmitting pathogens to the consumers of the fish.

**Figure 3.17. A large aquaculture facility on the Snake
River.**
Photo: Visuals Unlimited/Gary Will

To maximize yield, fish population densities in aquaculture operations tend to be high, leading to the generation of concentrated amounts of fecal matter and to the accumulation of uneaten food and dead fish. About 80 percent of the nitrogen contained in the food used in aquaculture operations ultimately ends up as waste, and much of the waste is released into surrounding waters. The BOD of the organic waste produced by one large marine aquaculture operation may surpass that of the untreated sewage from a town of 50,000 people. As described earlier in this chapter, excess organic waste released into a body of water will often result in eutrophication and death of native species. Localized dead zones are known to form beneath pens in salmon farms and unless the waste is dispersed by currents, the impact of the dead zones may extend a hundred feet or more beyond the pens.

Well managed fresh water aquaculture carried out in isolated ponds or tanks tends to have a less severe impact on the environment. Such facilities are usually separated from natural waterways. Also, pond drainage can be treated before release by passage through constructed wetlands, retention ponds or other treatment facilities.

To prevent loss of fish and shellfish by parasites or infectious diseases, aquaculture operators often have to use various parasiticides, antibiotics and other toxic chemicals. In addition, use of hormones, vitamins and other food supplements is common. These substances are usually added to the growing pen water directly or included in the food. Eventually excess chemicals make their way into surrounding ecosystems and may impact native species. As discussed in Chapter 2, the release of antibiotics into the environment encourages the development of antibiotic resistant bacteria, and aquatic habitats often harbor human pathogens for diseases such as salmonellosis, dysentery and cholera. There is also the question of whether humans would be affected by consuming fish may had been treated with these various chemicals.

Heavy population densities in fish farms also can result in the rapid spread of fish pathogens and parasites. Since mariculture operations often are located near natural habitats, the risk of transmission of infectious diseases and parasites from farm fish to native species is significant. There are numerous records of losses of native salmon, shrimp, oysters and other species through the introduction into the surrounding waters of parasites or microbial pathogens that originated in fish farms. For example, fish lice released by salmon farms in the Pacific Northwest are thought to be responsible for the loss of great numbers of wild salmon.

There are several things operators of fish farms can do to reduce risk of infectious diseases in their fish and their spread to native species. These include using only pathogen-free brood stock (see box below), reducing population densities, and in the case of pond- or tank-grown fish, disinfecting or filtering the effluent before discarding it.

Pasteur On Vertical Transmission

One of Louis Pasteur's greatest contributions to microbiology was the recognition that certain infectious diseases could be transmitted in animals down through generations by so-called **vertical transmission**. He accomplished that with his studies of the diseases of silk worms, and in so doing, saved the four century old

---

French silk industry from disaster. In 1865 Pasteur was asked to investigate why France's silk worms were dying. After 4 years of work, Pasteur reported that the worms had been infected with at least two pathogens that apparently were being passed down from previous generations through the eggs of the female silk moth. He showed the silk producers how to avoid problems by recognizing infected silk worm eggs through microscopic inspection. Thus by raising only pathogen-free worms, the French silk industry soon recovered. Pasteur gained greater hero status and the use of pathogen-free brood stock became a universal principle in many areas of animal husbandry. However, the most significant outcome of Pasteur's silk worm research was that it laid the foundation for the germ theory of disease.

Legislative Control of Water Pollution

Legislative control of pollution in our recreational waters is an enormously complex issue both technologically and politically. At the federal level, numerous laws have been passed to address the subject. The most significant legislation in this area is the **Clean Water Act (CWA)**, passed in 1972, with a number of subsequent amendments. It establishes a regulatory process to control point (known) sources of pollution and provides funds for upgrading water treatment plants. The act also mandates management practices to control non-point sources of pollution. Non-point sources mean sources that are not specifically identifiable, such as open space, streets and highways. Most significantly, the CWA made it possible for ordinary citizens to take polluters to court.

The **National Pollutant Discharge Elimination System (NPDES)** is a major feature of the CWA. The NPDES requires that anyone who discharges pollutants from a point source into "waters of the United States" must have a permit. Soil, rocks, sand, solid waste of any kind, discarded equipment and autos, sewage, chemicals, fecal coliforms, even hot water, are considered pollutants under the NPDES. Waters of the United States means practically any body of water, including the oceans out 200 miles, that is used for navigation, fishing and other recreational and industrial activities involving interstate commerce or travel. Significantly, wetlands also are included in this law. This basic law was passed simply to control the practice of unlimited dumping of materials of any type into the nearest body of water as a cheap way of getting rid of them. The NPDES provisions do not prevent the dumping of pollutants, they merely set limits on the quantity of given pollutants that are to be discharged to assure the discharge does not pose a threat to wildlife and humans.

The control of pollution from urban runoff can be approached at several levels. By definition urban runoff begins at homes, businesses and factories. Vigorous educational programs have been launched by governmental agencies to inform the public of the impact to the environment of over-irrigating landscaping and discarding hazardous materials into the flood control system (Figure 3.14). This approach is further supported in many communities by providing convenient household hazardous waste collections either curb-side or at central locations and encouraging used motor oil recycling. Frequent street sweeping has been shown to prevent tons of material from being washed into a community's flood control system.

Direct legislative control of urban runoff has been difficult because the runoff often originates from non-point sources such as parking lots, streets and so on. However, once it is collected in storm drains and empties into waters of the United States, the runoff is considered coming from a point source and falls under the NPDES. Thus counties, municipalities and other dischargers of storm water must have a NPDES permit and comply with CWA regulations, which include provisions for reducing pollutants and monitoring and reporting levels of pollutants to confirm that NPDES standards are not exceeded .

Section 303(d) of the Clean Water Act requires states to develop lists of bodies of water, referred to as **impaired waters**, that exceed limits for specific pollutants such as heavy metals, silt and pathogens. It has been estimated that nearly half of the nation's lakes, bays and river segments representing over 20,000 sites exceed the standards, about a third of which are impaired due to polluted runoff. Once impaired waters are identified, the next step is to determine the sources and quantities of the pollutants and then calculate the maximum amount of individual pollutant(s) from all sources that the body of water can receive per day and still meet water quality standards. That amount is known as the **Total Maximum Daily Load** or **TMDL**. Finally, states must describe plans for implementing controls on pollutant sources that would ensure water quality standards would be met to maintain the safety of the water, whether it is used as a source of drinking water or for swimming, fishing or other activities.

The **Beaches Environmental Assessment and Coastal Health (BEACH)** Act, signed into law in 2000, singles out the protection of beach water through programs for monitoring water quality and alerting the public when unsafe conditions exist.

The **Safe Drinking Water Act (SDWA)** of 1974 and its subsequent amendments was a major move by congress to assure United States citizens of safe water in their homes. The SDWA sets limitations for several dozen chemical pollutants that frequently occur in drinking water, as well as for turbidity and total coliforms. One section of the act, known as the Total Coliform Rule, sets specific limits for *E. coli* in water supplies designated for drinking purposes. The act also sets procedures for treating, disinfecting and analyzing water supplies and requirements for reporting to the appropriate authorities and the public when limits are exceeded. Finally, the act provides various mechanisms such as loans, grants and technical support to assist water suppliers in upgrading their facilities to comply with regulations.

In spite of these federal regulations concerning safe drinking water, much of the United States population is still at risk. One fifth of the nation's groundwater reportedly is contaminated with human enteric viruses, and only half of community water systems in the United States disinfect their water; in rural areas, practically none do. For individual homeowners who depend on private wells for their water, their only protection is their own vigilance by having their water tested periodically. The need for periodic testing is borne out by the fact that chemical storage tanks and sewer pipes can leak at any time and without warning.

Beaches in the United States experienced a 19 percent increase in advisories and closures due to high bacterial contamination in 2001 compared to 2000. Southern California won the distinction of ordering the most beach closures in the nation. The impact of polluted waters on our society is significant. Recreational activities that depend on clean water directly contribute about $50 billion dollars a year to the United States economy, and commercial fishing and shell fishing generate about an equal amount. In water-oriented communities clean water attracts a strong tourist economy and maintains high property values, assuring local governments of a stable tax base. A government study showed that the price of a house may be boosted by as much as 28 percent if it is within 300 feet of a body of water.

More importantly, there are the health effects of polluted water. The Centers for Disease Control and Prevention estimates that in the United States we can expect one million illnesses and approximately 900 deaths each year due to contaminated water. The World Health Organization has estimated that 3.4 million deaths occur annually worldwide as a result of unsanitary water supplies. It is clear there is still a lot to be done.

Further explorations:

Costerton, J. W. and P. S. Stewart. "Battling Biofilms." Scientific American July 2001: 74-81,

"Safeguarding Our Water." Scientific American Feb. 2001: 38+.

Toxic and Harmful Algal Blooms. 18 Mar. 2003 <http://www.bigelow.org/hab/impact.html>

UN Atlas of the Oceans. 18 Mar. 2003 <http://www.oceanatlas.org>

United States Environmental Protection Agency. 25 Years of the Safe Drinking Water Act: History and Trends. Washington: United States Environmental Protection Agency; EPA 816-R-99-0071999. 18 Mar. 2003 <http://www.epa.gov/safewater/sdwa/trends.html>

United States Environmental Protection Agency. Constructed Wetlands for Wastewater Treatment and Wildlife Habitat. Washington: United States Environmental Protection Agency; EPA 832-R-93-0051993. 18 Mar. 2003 <http://www.epa.gov/ owow/wetlands/construc/>

United States Environmental Protection Agency. <u>Introduction to TMDLs</u>. 18 Mar. 2003 <http://www.epa.gov/owow/tmdl/intro.html>

United States Environmental Protection Agency. <u>Wastewater Month.</u> 22 Mar. 2003. <http://www.epa.gov/npdes/wastewatermonth>

Chapter 4

Air: An Uninhabitable Habitat

Spontaneous Generation

As the ancient story goes, if one were to place a piece of raw meat in the open air, within a few days it would be heavy with maggots. Where did the maggots come from? None had been seen anywhere near the meat, and yet almost overnight they seemed to appear. The only possible explanation was that they arose spontaneously from the meat. The same explanation followed the observation that a cup of clear broth would soon become turbid with microorganisms. Clearly with the combination of the broth's organic ingredients, air and a "vegetative" or "life" force that was free to enter the cup, the microorganisms could be created spontaneously. The concept of **spontaneous generation,** as this explanation was called, prevailed for millennia until scientists began to devise experiments to test it.

One such experiment was carried out by an Italian physician, Francesco Redi, in 1668. He placed raw meat into two jars, covering one with fine gauze. Several days later the uncovered meat was riddled with maggots, but the covered one was not. Not surprisingly, the meat had attracted house flies, Redi reported, but they could not reach the covered meat because of the gauze. Redi discovered that maggots were part of the normal life cycle of flies, and that the flies had laid eggs in the uncovered meat and the eggs hatched. Eventually the stage of fly development we call maggots appeared. Apparently no one had ever connected maggots with flies before, and thus spontaneous generation at least as it related to the origins of maggots was disproved. Over the next two centuries several similar experiments followed that refuted other supposed instances of spontaneous generation.

The example of the clear broth becoming cloudy with microbes was probably the last remnant of the theory of spontaneous generation to be dealt with, and by none other than the prominent 19th century chemist-turned-microbiologist Louis Pasteur. Through the use of a variety of flasks with and without long necks (Figure 4.1), Pasteur proposed that the microorganisms that appeared seemingly spontaneously in the broth had simply dropped in from the air on dust particles.

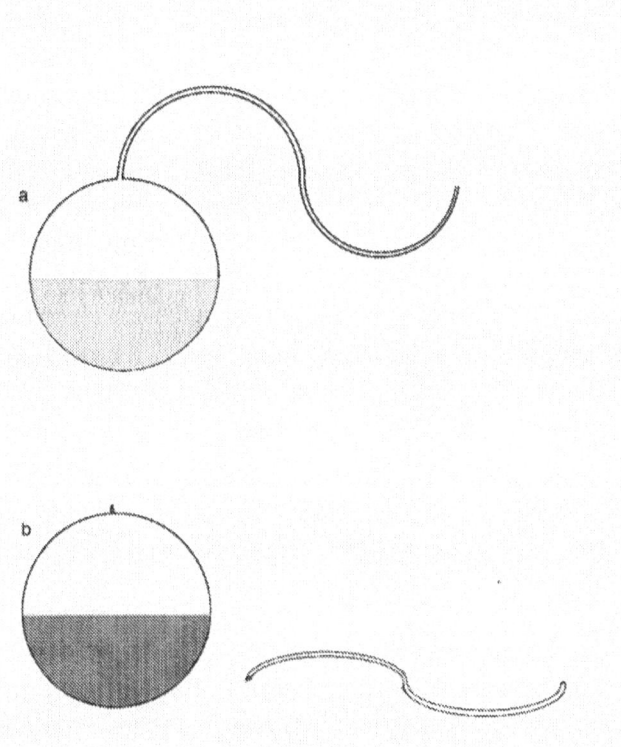

Figure 4.1 Pasteur's flasks
Pasteur's flasks he used to disprove spontaneous generation of bacteria in broth. Broth in flask (a) remained clear while the broth in flask (b) became cloudy with bacterial growth soon after its neck was broken off, which allowed bacteria-laden dust to enter the flask.

A few years later physicist Robert Tyndall devised a visual demonstration of dust particles falling into the broth, further supporting Pasteur's explanation. From these modest experiments the field of aerobiology was launched and at the same time the idea of spontaneous generation suffered a "mortal blow", as Pasteur put it. Oddly, even though the widespread existence of microorganisms had been established since the invention of the microscope in the 17th century, their presence in the air seemed largely to have been overlooked until Pasteur's time.

Spontaneous Generation Lives

One of the great paradoxes of biology has to do with theories of the origin of life on Earth. Fossil evidence, though recently challenged, suggests life first appeared within about one billion years after our planet was formed. The life either came from an extraterrestrial source or originated on Earth. If life started on Earth, scientists believe its development began with a few simple atmospheric gases that with the help of lightening, solar ultraviolet light and other energy sources, formed an array of organic chemicals and continued through a progression of molecules of increasing complexity to self replicating ones to ones that began to form organized structures that eventually became cells. Most biologists would agree that is a clear example of spontaneous generation.

Aerobiology

Aerobiology can be defined as the study of airborne biological material. That simple statement actually covers a multitude of fields, including epidemiology, medicine, agriculture, warfare, industry and space exploration. Biological material can include viable microorganisms, their metabolic products, or manufactured biologically active substances. There are numerous examples of human activities where aerobiology is involved. Here are a few:

•For many years aerobiology has been occupied
with the study of human airborne infectious diseases
such as tuberculosis and influenza.

•With the introduction of more complex surgical operations, such as hip replacements and open heart surgery, the surgical site remains exposed to the air for many hours, increasing the chance of infection by airborne bacteria.

•Losses of agricultural crops and farm animals through the spread of airborne disease has become a billion dollar problem.

•Aerobiology became a subject of modern military tactics when so-called germ warfare appeared a reality. More recent events have re-stimulated interest in the subject.

•The food, pharmaceutical and medical device manufacturing and supply industries had to find ways of producing and packaging products with minimum microbial contamination in a microbe-laden environment.

•Space vehicles that are designed to search for life on other planets must be assembled in near-microbe-free facilities in order to reduce the possibility of contaminating the planets with Earth microorganisms.

Microorganisms can become airborne through a wide variety of mechanisms. Here are a few actual scenarios:

•Influenza viruses may be ejected by a sneeze to be inhaled by an unfortunate individual a few feet away.

•Fungal spores on diseased plants may be swept up by the wind and carried many miles to infect another field.

•Water from a cooling tower that is contaminated with pathogens becomes aerosolized and drawn into a building's air conditioning system, distributing the pathogens to all parts of the building, leading to many illnesses and some deaths.

•An accidental spill in a microbiology research laboratory releases pathogenic bacteria into the air, infecting several workers.

•Loosely sealed envelopes containing anthrax spores were sent through the mail, infecting several individuals and killing five.

The air is not a hospitable environment for microorganisms. There are no nutrients present, and sunlight and the drying effect of heat and air currents can be lethal. In spite of these dangers, moderate numbers of viable microorganisms usually can be found in the air, particularly indoor air. While bacterial and fungal spores may survive for long periods, individual vegetative microorganisms may not remain viable for more than a few minutes after their release. However, those airborne microbes that have formed clumps or are associated with dust particles (See box below) or droplets of saliva (as in the case of the influenza viruses that are emitted by a sneeze) are more likely to withstand the air's hostile conditions and may survive up to several days.

Dusty Times

Over two billion tons of dust may be carried each year worldwide by winds. Satellite observations have revealed what appears to be clouds of dust being transported from Africa to the Americas, Europe and the Middle East. Dust from storms in China can be seen reaching eastward across North America all the way to Europe. Traces of mercury that have been detected in California air are thought to have originated from coal fires in China. Microbiological analyses of the dust that originated in Africa and reached the Caribbean have detected plant pollen as well as viable bacteria and fungi, some of which are known plant and human pathogens. Virus-like particles have also been detected in the dust, but they have not been positively identified.

Health workers in Southern California have reported that bacteriological counts go up significantly in coastal waters following strong off-shore winds ("Santa Anas"). At times the counts have been high enough to require health warnings to be posted. It is thought that the winds carry bacteria-laden dust from dairies and agricultural activities 50 to 100 miles inland from the coast.

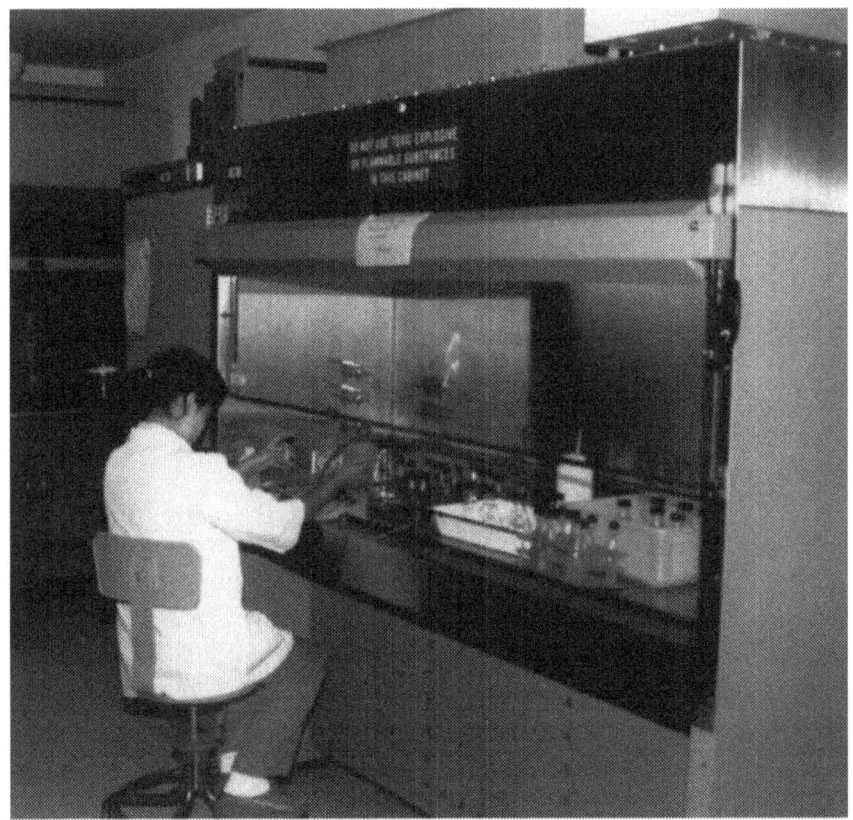

Figure 4.2. A unidirectional flow safety cabinet
in a microbiology research laboratory. Clean, filtered air flows down
onto the work surface from above, while room air enters the cabinet
through the narrow opening, protecting the worker from potentially
hazardous organisms escaping the cabinet.

Phases of Airborne Microorganisms
The life of an airborne microbe passes through three phases:
launch, transport and deposition.

As related above, microorganisms can be launched, or released
into the air by a number of mechanisms. In the case of microbes
associated with the human body, tests have revealed that a typical
human adult, depending on the type of clothing worn and other
factors, can shed from hundreds to tens of thousands of bacteria per

minute into the surrounding air. A single unprotected cough or sneeze can eject as many as 100,000 bacteria. Obviously these sources of airborne contamination are nearly impossible to avoid short of requiring everyone to wear particle-proof masks and clothing.

Workers in microbiology laboratories in hospitals, educational institutions and industrial facilities occasionally become infected with the microbes they are handling. The majority of these infections are thought to occur as the result of pathogens becoming airborne not by accidents but through routine laboratory activities. These incidents can be prevented by careful techniques on the part of the microbiologists and by using so-called biohazard cabinets (Figure 4.2) in which a gentle flow of air sweeps away any airborne microbes that may have been released and traps them in filters.

Outside of microbiology laboratories, routine, seemingly harmless practices can sometimes turn lethal. Prior to the anthrax incidences of 2001, post offices had been cleaning their equipment with jets of compressed air. The practice was immediately stopped when it was known that anthrax-containing letters had passed through sorting machines of a postal facility in New Jersey and the air jets were probably responsible for launching the spores into the surrounding air.

Once launched, the transport or dispersal of airborne microorganisms depends on ambient air currents. In the outdoors air movement is primarily produced by wind. Indoor air is subject to a variety of forces: normal room ventilation, drafts from open windows and doors, currents created by the movement of people and convection currents caused by heat sources.

Since moving air reduces chances of airborne microorganisms being inhaled or settling onto surfaces by gravity, indoor spaces with good ventilation are considered safer aerobiologically than ones with still air. This concept has been carried to the ultimate degree in the use of **unidirectional** (sometimes called **laminar flow**) ventilation in many industrial and medical facilities. Instead of delivering air into a room from a small, conventional vent (Figure 4.3a), the supply vent may be an entire wall through which filtered air is passed (Figure 4.3b). The special filters used in these facilities, known as **HEPA (High Efficiency Particulate Air)** filters, capture at least 99.999 percent of particles 0.3 micrometers and larger. Other types of filters, called **ULPA (Ultra Low Penetration Air)** filters, capture even greater numbers of particles. Since most microorganisms are larger than 0.3 micrometers, the air entering the room is nearly microbe free. Airborne microorganisms released in the room by people or activities are swept out by the air that is moving at about one mile per hour. The air exits the room through a grillwork that occupies the entire wall opposite where the air entered. The facility just described is referred to as having horizontal unidirectional flow. Some facilities are provided with vertical flow in which the ceiling acts as the supply vent and the entire floor, consisting of grillwork, the exit vent. Thus in such facilities the third phase of airborne microbes, deposition, is greatly reduced.

Figure 4.3. Airflow in a conventional (a) and a unidirectional (b) hospital operating room.
In (a) air enters the room and forms random eddies that pick up dust and microbes from the rooms and those emitted by personnel. Dust and microbes are deposited at other locations in the room. (b) Filtered air enters the unidirectional room through a bank of HEPA filters on the right. Moving at around one MPH, the air picks up any

infectious particles released by the surgical personnel and sweeps the particles out of the room before they settle onto the surgical wound.

Unidirectional air ventilation is currently being used in a number of industries that require ultra-clean air. Such research or manufacturing facilities, referred to as **cleanrooms**, may consist of an enclosed space with an area anywhere from less than 100 square feet to several thousand square feet or more. For instance, in modern microelectronic circuitry manufacturing, any dust contamination can lead to disaster. A solitary microscopic dust particle can short-circuit closely spaced conductors of a single component, resulting in the failure of an entire module. The manufacture of such devices requires it to be carried out in cleanrooms. Pharmaceutical companies and manufacturers of foods and medical devices are primarily troubled with microbiological contamination. As mentioned earlier in this chapter, these industries must find ways of producing products with low or zero microbiological contamination in a world that is pervaded with microorganisms. The manufacture and packaging of their products under unidirectional ventilation allows them to achieve that goal.

The assembly of spacecraft also is carried out in facilities with unidirectional ventilation. This is necessary not only to reduce particulate contamination that could cause a catastrophic failure, but in the case of unmanned planetary landers, to prevent microbial contamination of the landing site. One of the missions of these unmanned spacecraft is to look for signs of extraterrestrial life, not *E. coli* off the hands of a spacecraft assembler. Unidirectional air flow can be found in many hospital operating rooms, resulting in fewer infections due to the deposition of airborne microorganisms onto the surgical wound.. Unidirectional ventilation also is found in burn wards. Loss of protective skin in severe burn patients leaves them open to invasion by airborne bacteria. Unidirectional air ventilation greatly reduces chances of infections in such patents.

Infectious Irony

While unidirectional air flow in hospitals can reduce many types of infections, it is a surprising piece of irony that each year as many as 5 million people develop infections as patients in U. S. hospitals, resulting in nearly 90,000 deaths. In fact, such infections are so common that they have a name, **nosocomial** (Greek for *hospital disease*). There are many possible sources of such infections: visitors, attending physicians and nurses, other staff personnel, and food. Nosocomial infections are difficult to avoid since hospital patients are frequently more susceptible to infection due to the effects of chemotherapy, surgeries and having catheters and other devices implanted into the body.

David M. Carlberg, Ph.D.

The Air as a Weapon: Biowarfare and Bioterrorism

So far we have concentrated on natural forces that are responsible for microorganisms becoming airborne. Let's look at situations in which microorganisms are deliberately launched into the air as an act of war or terrorism.

Not all warfare involves bullets and bombs. Toxic chemicals and biological agents have long been part of the weaponry of nations, dissident groups and individuals. The use of poison gases in World War I is well documented, and in prior centuries of human history accounts have appeared telling of military forces exposing their enemies to highly infectious material such as plague-infected cadavers or blankets and clothing that had been used by smallpox victims. For the most part, these events occurred centuries before the true cause of infectious diseases was understood. Renewed interest in the development of biological weapons began to appear in the 1930s and accelerated during World War II. The United States, the U.S.S.R. and the United Kingdom as well as Japan and Germany established extensive research and manufacturing facilities for both chemical and biological weapons. The end of WW II did not end the biological arms race; it appears to have continued into the 21st century.

In 1972 the U. S., the USSR and nearly 140 other nations signed the Biological and Toxin Weapons Convention, pledging to end development of offensive biological and chemical weapons and to destroy stockpiles of such weapons. Unfortunately the treaty lacked any meaningful provisions for verification and by 1989 at least 10 signatory nations, including the USSR and Iraq, covertly continued developing and manufacturing biological weapons. By 1995 the list had grown to 17 nations with bioweapon capability. An unusual outbreak of anthrax in the Soviet Union (see below) in 1979 was the beginning of a series of revelations that eventually led to the admission by the Russian government of extensive bioweapons development. The United Nations has taken on the responsibility of confirming adherence to the convention. For example, Iraq's bioweapon development was exposed through on-site investigations by United Nations inspectors during the 1990s. While the recent Iraqi regime has fallen, there is little evidence that bioweapon development has ended in any of the other countries with bioweapon capabilities.

An Accidental Bioattack

Over a period of several days in the spring of 1979, hospitals in the Soviet Union city of Sverdlovsk (now called Yekaterinburg) suddenly found themselves with several hundred patients exhibiting signs and symptoms of a severe respiratory infection. At least 66 died. Autopsy results clearly pointed to inhalation anthrax as the cause of death. Nearly all the patients lived or worked within a few miles of a secret Soviet military biological warfare facility

Soviet government officials desperately tried to hide the incident, since the facility was operating in defiance of the 1972 Biological and Toxin Weapons Convention that the Soviet government had signed. Initially the government blamed the outbreak on contaminated meat, an obviously illogical and misleading conclusion in light of the autopsy results. It was reported that Soviet scientists even delivered papers at a conference in Washington with faked medical evidence supporting the contaminated meat explanation. But after the fall of the Soviet government, world reaction eventually forced officials of the newly formed Russian government to admit that an accident at a biological weapons development facility caused the release of a cloud of anthrax spores into the air.

A maintenance worker had failed to replace a bacteria-proof filter on a tank for drying anthrax spores and an increase in internal pressure apparently ejected some of the bacteria into the air. By checking weather data, it was shown that the sickened individuals were exactly downwind of the facility at the time of the incident. The amount of bacteria that was released has not been confirmed, but it probably was not large, since only a relatively small fraction of the city's population was affected. Fortunately, because the incident occurred at night, students and teachers at schools close to the facility were spared. In fact, there were no fatalities under the age of 24. Over 70 percent of the fatalities were men, due primarily to the close proximity to the bioweapons facility of a ceramics manufacturing plant that was operating a night shift at the time of the accident.

A Modern Day Bioattack

A biological warfare attack in modern times might involve the release into the air of a highly infectious agent that would become dispersed amongst military or civilian personnel. The air becomes a weapon. Alternatively a biological agent might be placed covertly in a food or water supply. The goal is always the same: to cause death, illness or at the least panic within the target population, weakening its ability to defend itself or creating political, economic and social instability.

The choice of agent depends on several factors, such as whether the attack is intended to cause widespread death or merely to incapacitate a population. The standard list of lethal agents include those bacteria that cause anthrax, plague, and tularemia, and the viral agents for smallpox and hemorrhagic fever. Botulism toxin and *Staphylococcus* enterotoxin are also on many lists. Properly disseminated, virulent strains of the enteric bacteria *Salmonella*, *Shigella* and *E. coli* can be expected to disable significant numbers of susceptible military or civilian personnel. The manner of delivery would depend on whether the agent is a respiratory or an intestinal pathogen.

A biological weapon need not be directed at human populations. The destruction of crops and farm animals with specific plant and animal pathogens can be a devastating blow to a targeted population. As a case in point, Iraq's bioweaponry was reported to include such agents as wheat smut fungus and camelpox virus.

Biological weapons have a number of advantages over conventional weapons.

•Unless effective monitoring devices are operating, an assault with a biological weapon initially may go unnoticed. A population may not be aware it had been under attack until signs and symptoms of disease begin to appear, perhaps days or weeks after the attack, too late perhaps to apply any preventative measures. If medical personal are unfamiliar with the particular agent, additional delay in identifying the agent and treating the disease is certain. Confusion may follow. However, efforts are underway to inform physicians of the signs and symptoms of the diseases most likely to be associated with a bioattack. Few physicians practicing in the U. S. today, for example, have ever seen anyone in person with the signs and symptoms of Ebola hemorrhagic fever, smallpox or anthrax.

Even if the agent is identified immediately, it may not be clear in the beginning whether the cases are the results of a biological attack or a natural outbreak. For instance, were the recent, sudden outbreaks on several cruise ships of intestinal disease the result of natural infections or deliberate terrorist acts? Public health officials agreed that the nearly simultaneous incidences were natural infections caused by a group of common and highly infectious viruses known as **Norwalk-like viruses** or **Noroviruses**. What about the outbreaks of severe acute respiratory syndrome (SARS) that affected hundreds of air travelers, some fatally, in 2003? The nation-wide outbreaks of infections by the mosquito-borne West Nile virus is another example. West Nile virus is well known in Africa and Asia but was unknown in the U. S. until 1999. By 2002 it had spread throughout most of the United States. West Nile viral encephalitis is considered one of many so called Emerging Infectious Diseases (see Chapter 1) that are spread through populations by natural mechanisms.

•Biological weapons are inexpensive and easily manufactured compared to conventional weapons. Many agents can be prepared as easily as brewing a batch of home-made beer, although considerably riskier. Special precautions must be practiced by the perpetrators to avoid exposing themselves to the agent. Such precautions are taught in every college introductory microbiology course. To increase their infectivity, biological agents frequently must be "weaponized." Weaponizing biological agents usually does require specialized equipment and additional knowledge and skill and involves an even greater level of risk for the preparers. Agents intended for air dispersal must be prepared with biological and physical properties to withstand the inhospitable conditions of being airborne, to remain airborne long enough to reach the intended target, and if a respiratory agent, to exhibit a size range that would penetrate the lungs and avoid the body's protective mechanisms.

•Biological weapons are self-propagating. A small amount of a highly infectious agent can spread quickly through a susceptible population.

Biological weapons have certain disadvantages that may limit their usefulness: For one, they can backfire. Japan is reported to have released plague-infected fleas against Chinese forces in Manchuria in 1942, which resulted in a number of Japanese solders becoming infected. On a larger scale, a biowarfare attack aimed at one part of the world could spread to other regions, affecting unintended populations, including the perpetrators, and even becoming a worldwide, catastrophic **pandemic.** A pandemic is an epidemic that crosses international boundaries.

Since the list of agents most likely to be used for preparing biological weapons is relatively short, populations such as combat soldiers can be vaccinated against those agents for which vaccines are available. Some biowarfare agents are susceptible to many common antibiotics. There is, however, the disturbing possibility that an agent far down the list, or not on the list at all, is used. In addition, there has been speculation that through genetic engineering it may be possible to breed strains of, for example, anthrax bacteria that would be resistant to the common anthrax vaccines or antibiotics.

Recent attacks using biological agents have been of limited scope, carried out by terrorist groups or individuals. In 1984, in an attempt to influence a local election, a religious cult in Oregon deliberately contaminated salads in restaurants with *Salmonella* and caused illnesses in over 750 people. Fortunately there were no deaths. In 2001, five deaths and several illnesses from anthrax were attributed to exposure to envelopes containing *Bacillus anthracis* spores that had been mailed to specific individuals. Apparently the envelopes were not fully sealed, allowing the spores to escape while en route to the addressees. The spores become aerosolized and infected several presumably unintended targets. Most of the victims were mail handlers.

Sick Building Syndrome

Certain groups of microorganisms, particularly molds, can trigger allergic and other adverse reactions in humans. As many as 100 species of mold have been identified as potentially causing health problems related to unhealthful air. Molds can grow nearly anywhere, such as on walls, books, beneath carpeting and on ventilation filters. Molds can grow where there is very limited moisture and nutrients, but areas hit by water damage are particularly susceptible to mold growth. Mold growth produces enormous numbers of spores as well as toxins that become airborne and disseminated throughout a building. On contact with building inhabitants the fungal products can cause a variety of signs and symptoms, including runny nose, itchy and watery eyes, headache and difficulty in breathing.

Office buildings, shopping malls and residences are sometimes struck by what is referred to as the **Sick Building Syndrome** in which occupants complain of one or more of the above symptoms. Often sick building syndrome has been shown to be caused by mold growth somewhere in the building, and air sampling techniques to detect those fungi commonly associated with allergies usually can trace the source of the problem. Chemicals, dust and noise pollution have also been identified as possible causes of sick building syndrome.

Determining Airborne Microbial Populations

There are several reasons to measure numbers of airborne microorganisms: to determine if a ventilation system is working properly, to trace the source of an outbreak of an airborne infection and to determine possible causes of sick building syndrome.

The best way to determine if a ventilation system is operating as designed is to measure the numbers of airborne dust particles or viable bacteria in the facility. A number of electronic counters are available that can deliver instantaneous counts of airborne particles, but enumerating airborne microorganisms is another matter. The electronic counters cannot distinguish microorganisms from inert dust particles. Airborne microbes must be captured and cultured in order to estimate their numbers. Several air samplers designed specifically to collect airborne microorganisms are manufactured. Most of these samplers operate on the principle of inertia and are known as **impacters**. Airborne microbes (usually attached to dust or other larger particles) are accelerated by drawing them into the sampler by a vacuum pump or turbine. The air flow is sharply deflected away from a collection surface. Airborne particles, because of their inertia, will continue on a more or less straight path and impact the collection surface and become fixed to the surface, which is often an agar growth medium. The medium is then incubated and the number of colonies that form is then related to the volume of air that had been sampled. Typical counts for a business office, for example, may range from 0.1 to 5 colony forming units (CFU) per cubic foot of air. Note that as in the case for waterborne counts, airborne microbial counts are also reported in terms of CFUs rather than individual cells. Also, as in water analysis, the time necessary to grow captured airborne microbes causes a delay of as much as 24 hours or more before results are known. Instruments that instantaneously detect airborne microorganisms have been under development for decades, primarily as devices for detecting biological warfare attacks, but they generally have suffered from lack of sensitivity and specificity. Some progress has been made in developing microbial detectors that deliver a result in less than one hour, but prior knowledge of the specific organism(s) being sought is usually necessary.

Microorganisms and Global Warming

When Earth was first formed about 4.6 billion years ago, its atmosphere was quite different from what it is today. It consisted mainly of carbon dioxide (CO_2), methane (CH_4), nitrogen (N_2), ammonia (NH3) and hydrogen (H2). Most importantly, it lacked any significant amounts of gaseous oxygen (O_2), although there were large amounts of oxygen in the planet's abundant water (which may have come from extraterrestrial sources some time after the planet was formed. See Chapter 5). While O_2 was continually being released through the disassociation of water vapor by solar radiation, it didn't remain in the atmosphere as O_2 for long. It quickly reacted with oxygen-hungry elements in Earth's freshly formed crust to become part of the planet's mineral composition.

After about a billion years the first signs of life appeared on our planet, and those primitive organisms must have been anaerobic, since there was still little gaseous oxygen present in the atmosphere. Within less than another billion years, photosynthesis appears to have evolved. Photosynthesis is the ability to convert light energy from the sun into chemical energy. From genetic evidence it appears that the ability to carry out photosynthesis was the result of horizontal gene transfer between several early forms of bacteria, the descendants of which are still widely distributed in nature. That gene sharing eventually resulted in the assemblage of photosynthesis genes into a single organism. The photosynthesis conducted by these organisms was not the type that we normally associate with modern plants, that is, photosynthesis that results in the release of oxygen. This more primitive form is referred to as anoxygenic photosynthesis: photosynthesis without releasing oxygen.

Oxygenic photosynthesis appears some time after, probably about 2.8 billion years ago, or perhaps earlier, and carried out by organisms similar to what we now call cyanobacteria. Oxygenic photosynthesis refers to the ability of organisms to capture energy from sunlight and at the same time break down H_2O to release O_2 as a gaseous by-product. At this point in time, levels of gaseous oxygen in our atmosphere began to increase. The eventual appearance of plants added to the buildup of gaseous oxygen, making possible the emergence of more complex organisms that could take advantage of the greater energy yield of aerobic respiration. It's been estimated that the appearance of oxygenic photosynthesis increased the organic productivity of life on Earth by two or three orders of magnitude or more. Atmospheric oxygen also led to the formation of a layer of ozone (O_3) that enclosed our planet with a protective shield against lethal solar ultraviolet radiation. Eventually oxygen accumulated in our atmosphere to reach its present level, about 21 percent.

Of the other atmospheric gases, nitrogen (N_2) is the most abundant at 78 percent, with carbon dioxide (CO_2) appearing at about 0.04 percent. The balance consists of the so-called trace gases, methane, argon, oxides of nitrogen and others. Carbon dioxide and the trace gases methane and nitrous oxide (as well as water vapor) are known as **greenhouse gases** because they are responsible for the **greenhouse effect**. As solar energy strikes Earth, much of it is retained as heat absorbed by the planet's surface, but some of the heat is re-radiated back into the atmosphere. Greenhouse gases in the atmosphere absorb the re-radiated heat and prevent it from escaping into outer space, somewhat like how the glass of a nursery greenhouse retains heat (Figure 4.4). The greenhouse effect is a natural phenomenon that is necessary for maintaining Earth's moderate temperature. If the atmosphere were not present and Earth lacked the greenhouse effect, our planet's maximum temperature would be about –20°F, far too cold for life as we know it to exist.

Figure 4.4. A typical nursery greenhouse.
The glass roof allows solar light and heat to enter the greenhouse, but prevents excess heat from escaping, which results in a build-up of heat within the greenhouse.

The concern of many scientists is that atmospheric greenhouse gases are increasing at an unprecedented rate. Atmospheric carbon dioxide concentrations are presently rising at a rate of about 0.4 percent per year, and methane is increasing annually by about 1 percent (Figure 4.5). Consequently, over the last 100 years carbon dioxide has increased about 30 percent, nitrous oxide 15 percent and methane has doubled in that period of time.

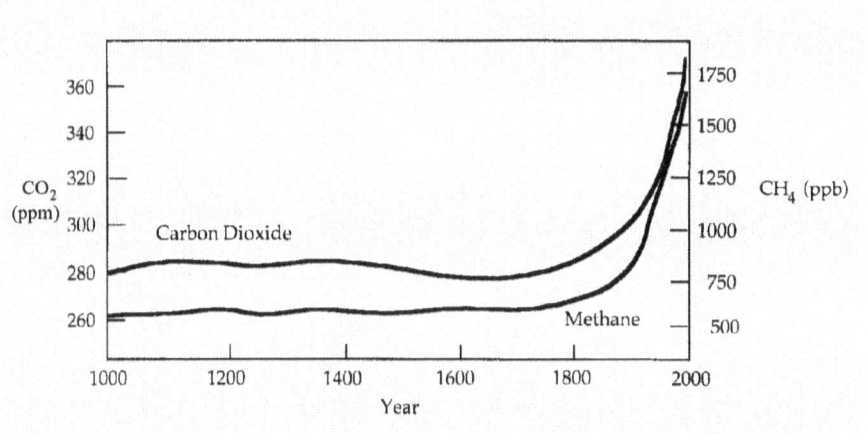

Figure 4.5. Global rise in atmospheric carbon dioxide and methane over the last 1000 years.
Early figures are based on geological evidence, later on measurements. Figure from USEPA.

It has been observed that Earth's mean temperature has been rising steadily at the same time the increase in greenhouse gases has been seen (Figure 4.6). Some scientists have estimated that this temperature rise, known as **global warming**, may reach from 3 to as much as 8°F over the next century. The impact of such a seemingly small temperature increment would be considerable. A rise in sea levels would be one of the most costly consequences of global warming. Due to melting polar ice and thermal expansion of the oceans, the rise would lead to flooding of coastal cities. Global warming would cause changes in weather patterns, severely impacting transportation and agricultural activities. Plant pathogens that are normally killed by winter temperatures would survive more mild winters, only to cause more severe outbreaks the following spring. In addition, plants appear more susceptible to disease following mild winters. From a public health viewpoint, the change could be disastrous. Tropical climates would expand northward, bringing with them to cities such as Los Angeles, Dallas, Miami, Rome and Hong Kong diseases such as malaria, yellow fever and dengue fever, which depend on hot, humid climates to sustain the insect vectors that transmit them. Wildlife would suffer a similar effect.

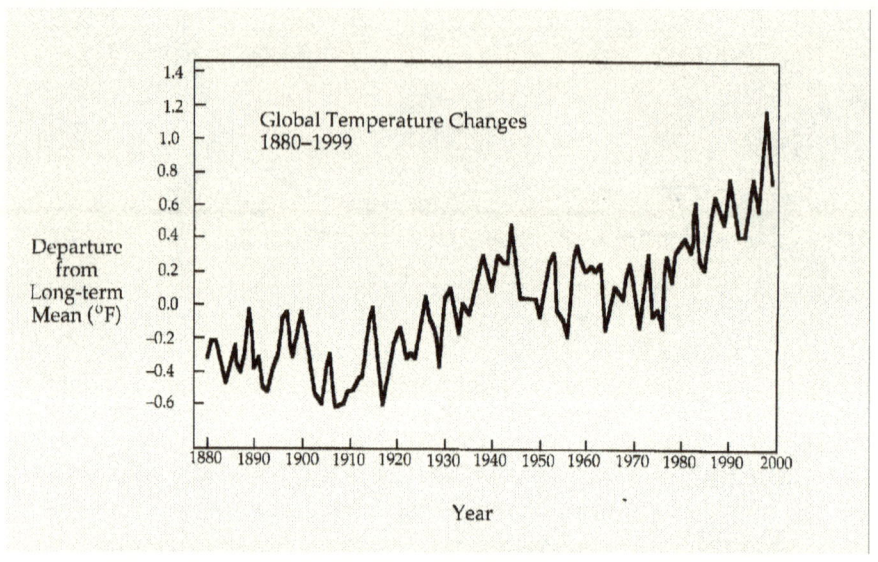

Figure 4.6. Changes in global temperatures since 1880. Figure from USEPA.

Critics frequently point to cold weather extremes such as the late winter 2003 snow storms that hit the Northeast U. S. as evidence against the existence of global warming. Many such extremes are seen in Figure 4.6, but these are normal short-term climate fluctuations and have little to do with the long-term warming trend that is clearly seen in the figure.

Is there a connection between the increased release of greenhouse gases and global warming? It is generally believed that there is. Most scientists also agree that human activities associated with the industrial revolution that began in the late 19th century, particularly the dependence on fossil fuels, is the major cause of the increase in greenhouse gases, mainly CO_2. For example, it has been estimated that about 98 percent of CO_2 emissions in the U. S. is due primarily to the burning of fossil fuels. The clearing of rain forests for farming is another human activity that leads to increases in atmospheric CO_2. That is because the trees of a forest, due to their concentrated biomass, are thought to represent a significant trap of atmospheric CO_2, whereas the relatively small biomass and short growing season of crop plants do not absorb as much CO_2. Global warming itself may trigger the release of even more CO_2 and methane. As soil, peat bogs and wetlands of northern latitudes are warmed, increased microbial activity may generate the release of greater amounts of these greenhouse gases.

Several synthetic gases, such as the chlorofluorohydrocarbons (CFCs), are used as refrigerants and aerosol can propellants frequently end up in the atmosphere. These gases are more powerful greenhouse gases than CO_2. For example, one such gas, CFC-12 or dichlorodifluoromethane, is over 8000 times as effective in trapping heat as is CO_2. CFCs also are thought to contribute to the destruction of our atmosphere's ozone layer, which protects us from the most hazardous wavelengths of solar ultraviolet radiation. Another gas that is used in many industrial applications, sulfur hexafluoride, is nearly 24,000 times as effective as a greenhouse gas than is CO_2. The use of these gases has been drastically curtailed in recent years in recognition of their effect on the environment.

Carbon dioxide is soluble in water. With over 78 percent of our planet covered with water, a lot of atmospheric CO_2 ends up in our oceans and lakes. If it were not for the capture of CO_2 by large bodies of water, atmospheric CO_2 concentrations would be even higher than they are. However, the capacity of the oceans to absorb CO_2 is not unlimited. The increased production of CO_2 by human activities appears to be overcoming this capacity, resulting in a significant increase in the amount of CO_2 in the atmosphere and a potential acceleration of global warming.

The CO_2 that is dissolved in the oceans is readily available to marine algae and photosynthetic bacteria to be converted to cell mass. These organisms, known as **primary producers**, are at the very base of the oceanic carbon cycle. In comparing primary productivity in various ecosystems (cornfields, wetlands, deserts), the open oceans are one of the poorest in terms of weight of organic matter produced per unit area per year. However, since over three-fourths of the planet is covered with oceans, their total contribution in converting CO_2 to cell mass and the release of O_2 is considerable. In fact, about half of

Earth's primary productivity, as well as half of its O_2 production, occurs in our oceans.

Studies have been launched to find ways of increasing primary productivity in the oceans to offset the rising levels of CO_2 in our atmosphere. One strategy to stimulate primary productivity is to increase concentrations of scarce nutrients like iron. Small pilot experiments in **ocean fertilization,** as it is called, have been encouraging, but doubts concerning whether the plan would work on a large scale and its impact on the ecology of the ocean have been strong deterrents for any widespread applications of this approach.

Through photosynthesis plants provide gaseous oxygen to our atmosphere, but even that process may have a microbial connection. As discussed in Chapter 1, chloroplasts are small membranous structures within plant cells that make it possible for plants to carry out photosynthesis. Chloroplasts (as well as mitochondria) contain DNA and are capable of independent growth and replication. Genetic analysis of chloroplasts has suggested that these structures at one time may have been free-living photosynthetic microorganisms that happened to parasitize some eucaryotic cells. In time the eucaryotic cells evolved into plants, carrying the organelles of photosynthesis that were once living organisms themselves.

Under anaerobic conditions carbon is emitted by certain bacteria as the gas methane (CH_4), also a greenhouse gas. Wetlands, rice paddies, sanitary landfills and oil extraction are some of the most common sources of atmospheric methane. In contrast, soil bacteria known as methanotrophs can utilize methane as a source of carbon. This aspect is covered in more detail in Chapter 5.

Microorganisms have played and continue to play a variety of roles related to the composition of our atmosphere. Some microorganisms consume O_2 and release CO_2 just as animals do, while others are more like plants, able to take up CO_2 and release O_2. Some bacteria produce methane while others are able to consume it. Besides their role in determining the composition of our air, microorganisms also use air as a means of transportation, even though it does not support their growth, and in most instances, is hostile to their survival.

Further Explorations:

"Feeling the Heat: Special Report on Global Warming." Time 9 Apr. 2001: 22+.

Griffin, Dale W., Kellogg, Christina A., Garrison, Virginia H. and Shinn, Eugene A. "The Global Transport of Dust." American Scientist May-June 2002: 228-235.

Hazen, Robert M. "Life's Rocky Start." Scientific American Apr. 2001: 77-85.

"Life in the Universe." Scientific American Oct. 1994.

Oak Ridge National Laboratory. Carbon Dioxide Information Analysis Center. "Frequently Asked Global Change Questions." 18 Mar. 2003 <http://cdiac.esd.ornl.gov/pns/faq.html>

Simpson, Sarah. "Questioning the Oldest Signs of Life". Scientific American Apr. 2003: 70-77.

Teirno, Philip M. Jr. Protect Yourself Against Bioterrorism. New York: Pocket Books, 2002.

United States Environmental Protection Agency. "Global Warming." 18 Mar. 2003 <http://www.epa.gov/globalwarming/climate/index.html>

Wolfe, David W. "Out of Thin Air." Natural History Sept. 2001.

Chapter 5

Earth: Where Life Begins and Ends

Biogeochemical Cycling

Some 4.6 billion years ago, perhaps 10 billion years after the birth of the Universe, a spinning nebula of gas and dust agglomerated into masses that eventually became the Sun and planets of our solar system. This process may have taken from 30 million to 100 million years, scientists estimate. The 90 or so chemical elements contained in the nebula had been formed in the intense nuclear heat of stars that had emerged at the time but no longer exist. The elements combined with one another to produce the over two thousand kinds of minerals that make up our planet's land masses, while other gaseous elements, such as nitrogen, methane and hydrogen together with the most abundant gas, CO_2, formed the planet's atmosphere. Traces of ammonia, sulfur dioxide and hydrogen chloride (gaseous hydrochloric acid!) made it a particularly noxious atmosphere, at least by present standards. Some of the elements that formed the land, like gold and osmium, appeared in minute amounts while others occurred in enormous abundance, such as silicon, aluminum and oxygen (not as a gas but combined with other elements in minerals or possibly water).

The role of microorganisms in establishing the physical and chemical properties of Earth's atmosphere and outer-most layer or mantle has long been underestimated, even ignored. Scientists are beginning to realize that microorganisms may have been the most important factor in shaping the chemical nature of our world as we now know it. We already described how our atmosphere had been altered from its original composition through the action of microorganisms. And as noted in Chapter 1, microorganisms can be found in nearly every niche on Earth, including ones uninhabitable by other forms of life. The microbes have not been passive tenants of these niches, however. Over the eons of their occupancy the microorganisms have used their enormous metabolic reach to transform the very geochemistry of Earth into what it is today. For instance, it's beginning to appear that caves, many of which have spectacular mineral formations such as the Carlsbad Caverns and Lechuguilla Cave in New Mexico and Mexico's Cueva de Villa Luz were actually formed (and in some cases are still being formed) by microbiological activity. In this chapter, we will look at just a few aspects of the influence microbes have on the chemical makeup of our planet.

Whence Water?

Some scientists believe that water was not part of Earth's original composition, but was deposited on the planet by water-logged comets or asteroids that struck it during the first few hundred million years of its existence. In addition, these heavenly water carriers also may have contained simple organic compounds that provided primitive building blocks for the eventual formation of life in Earth. Analysis of meteorites that have impacted Earth has revealed traces of such building blocks, including amino acids and sugars. But then, life itself may have been delivered to Earth by these missiles. Some scientists claim that a few recovered meteorites contain what appear to be fossilized remains of microbial cells, but that has been challenged by other scientists.

According to one widely held theory on the origins of life on Earth (described in more detail in Chapter 4), over a period of a billion years or so after Earth was formed a handful of chemical elements assembled into living organisms, each element perfectly suited for its particular role. While the exact chemical composition of those early bacteria-like organisms is unknown, we do know that 98 percent of the dry weight of a typical present-day bacterial cell is made up of just seven elements: carbon, oxygen, hydrogen, nitrogen, sulfur, phosphorus and potassium, the remaining two percent representing perhaps another dozen elements that are also necessary for the many functions of a living cell.

Organisms have lived and died on Earth for as much as three and half billion years. During that time the chemical elements that made up the microorganisms, plants and animals could have accumulated in their remains until some elements would have become depleted. Life then would have gradually faded away. Of course we recognize that did not happen. Dead organisms decompose through microbial activity, releasing the chemical elements that produced them to generate future organisms. While certain amounts of the elements are retained by the microbial decomposers for building more of their own cells, a significant portion of the elements is returned to the environment to be used by future generations. This complex web of material turnover is known as **biogeochemical cycling**.

The two most prominent elements that are involved in biogeochemical cycling are carbon and nitrogen. Carbon makes up half of the dry weight of a typical bacterial cell, and nitrogen, fourteen percent.

Much of Earth's biogeochemical cycling begins and ends in its soil. Soil is defined as the upper layer of the land surface of our planet, consisting of mineral particles and organic matter. It is the medium in which plants grow. Life on our planet depends largely on plants, hence life on Earth depends on the health of our soil. Healthy soil must contain all of the nutrients necessary for plant growth, except carbon dioxide, which plants get from the air. Soil becomes unhealthy for plants in many ways, such as lack of nutrients, too little or too much water, or the presence of toxic pollution.

The Carbon Cycle

Much of Earth's carbon is tied up in its crust as carbonate minerals and is generally inaccessible for use by living organisms. The most biologically accessible source of carbon is the carbon dioxide (CO_2) in our atmosphere. While CO_2 represents only 0.04 percent of the air by volume, it is the major player in the carbon cycle; it is the ultimate source of most of the carbon in all living organisms.

The CO_2 in the atmosphere is biologically accessible only to plants, algae and some bacteria. Animals must wait until plants have "fixed" the CO_2 into organic matter which animals can then consume for the energy to run their metabolism and the carbon to build cell structure. If plants were to stop processing CO_2, all animal life would eventually perish. By a series of biochemical steps plants take the CO_2 and use it to synthesize the sugar glucose, a molecule consisting of six carbon atoms, 12 hydrogen atoms and six oxygen atoms ($C_6H_{12}O_6$). The overall process from CO_2 to glucose is known as **photosynthesis**. Making glucose requires a lot of energy, which must be captured from sunlight by nature's first solar cells, called **chloroplasts,** found in the leaves of all plants. Then, with glucose as the starting material, a plant can begin to assemble the enormous variety of organic molecules that make up its roots, stems and leaves. For example, the basic structure of plant tissue consists of very large polymeric molecules called cellulose and hemicelluloses. **Cellulose**, the most abundant polymer on Earth, consists exclusively of glucose molecules arranged in rigid, straight chains of up to 10,000 subunits in length. These molecules are the equivalent of the steel beams that make up the skeleton of a high rise office building. Hemicelluloses are quite heterogeneous, consisting of a variety of sugars and other organic subunits arranged in very large branched chains, much like the mesh of reinforcing bars that are part of our office building. Pectin, used as a common thickening agent in certain foods, is an example of a hemicellulose. Starch is also a polymer of glucose and is the substance plants normally use to store energy and carbon. Since it is a relatively small molecule, glucose cannot be stored in cells by itself, for it would leak out of the cell and become lost. But in a huge polymer like starch, there is no danger of it getting out of the cell.

Like energy stored in a battery, much of the energy that went into the synthesis of the glucose is still bound up in the sugar molecules that are now part of a plant's structure. Thus whenever an organism consumes the plant, be it an animal, insect or microorganism, that energy stored in the glucose will be released and made available to that organism for its own metabolic needs.

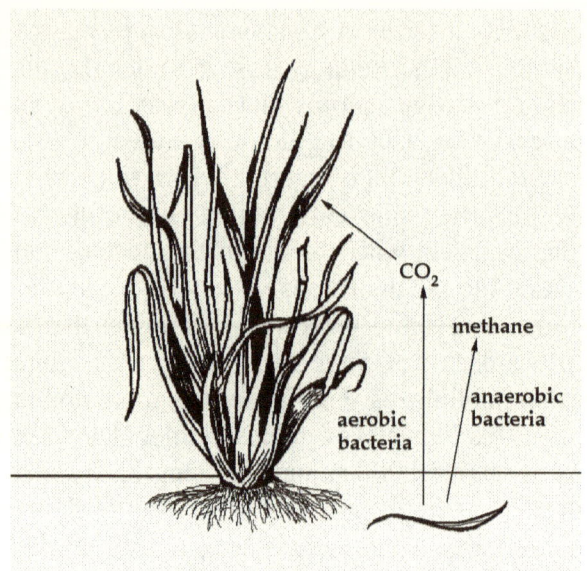

Figure 5.1. The carbon cycle.

The carbon from the organic compounds in a fallen leaf is utilized by soil microorganisms to make more microbial cells. Under aerobic conditions excess carbon is released as carbon dioxide, which is taken up by the plant to make more leaves. Under anaerobic conditions, much of the carbon is released as methane, a form of carbon that is unavailable to plants.

Consider a leaf that has fallen from a bush (Figure 5.1). If one were to observe the leaf over several days or weeks, it might seem gradually to disappear until there was practically no trace of it. Unless the leaf is devoured by insects or an herbivorous animal, most of the leaf has probably undergone decomposition by microorganisms in the soil. Since soil microorganisms do not have teeth, they cannot bite off pieces of the leaf. Instead, they release a variety of enzymes out of their cells that dissolve away the leaf's polymeric structure until all that is left are sugars and other small organic molecules, which are readily taken up by the microbial cells. The energy stored in the sugar molecules that the plant had originally captured from the sun can now be extracted by the microorganisms, and some of the sugars' carbon atoms can be utilized to build more cell structure. Eventually the microorganisms release excess carbon as CO_2. Once it is in the atmosphere, the CO_2 may be captured by a plant to make more leaves and the carbon cycle is complete. Of course a similar cycle occurs if an animal were to consume the leaf. Through the process of digestion and metabolism the carbon from the leaf, if not used to build more cell structure, is exhaled as CO_2.

Microbial Digestive Aids

While many animals, including humans, include plants in at least part of their diet, they cannot digest the major structural molecules of plants, the cellulose and hemicelluloses. They lack the necessary enzymes. While they may be able to extract the smaller organic molecules, such as sugars, along with the minerals and vitamins in the plant material, they are unable to take advantage of the great amount of carbon and energy that is stored in the polymers of glucose. The polymers pass through their digestive tracts essentially untouched, albeit their presence is supposed to provide some healthful benefits. And for those animals, known as herbivores, that depend entirely on plant-based food, the loss is even greater. They are forced to eat enormous quantities of plants just to maintain a normal level of nourishment.

However, by an astonishing circumstance, a particular group of herbivores known as **ruminants** have acquired the ability to digest cellulose due to the presence of certain bacteria and other microorganisms in their stomachs. Cattle, sheep, deer and goats are familiar examples of ruminants. The resident microbes produce cellulase and other enzymes that convert cellulose and other constituents such as pectin into significant quantities of nutrients for their animal hosts. This symbiotic activity takes place in the **rumen**, a chamber through which food passes before it reaches the stomach. Bacterial populations in the rumen can run to more than 10^{10} cells per gram of contents. Over 20 species of rumen bacteria have been identified in cattle. To aid the digestion of the cellulose and pectin, a ruminant will regurgitate its rumen's contents and engage in what is commonly known to all as chewing its cud.

In a similar example of mutualism, protozoa that live in the hindgut of termites make it possible for these insects to digest wood, which is mostly cellulose. Methane gas, a greenhouse gas (see below) is one of the major by-products of the microbial associations in both ruminants and termites.

Methanogenesis

The leaf that was decomposed by soil microorganisms met its fate in an aerobic environment. However, if that leaf had settled to the bottom of a bog or was buried in a landfill, its decomposition would have been under anaerobic conditions and would have proceeded much slower and with a different outcome. Some of the leaf's carbon may be released in the form of methane (CH_4) instead of CO_2. Methane is a flammable gas that is most often associated with petroleum and coal deposits. It is the major constituent of natural gas that is used for cooking and heating in many homes and businesses. As pointed out in Chapter 4, methane is produced by microorganisms in wetlands, rice paddies and other water-saturated soils, in solid waste landfills and anaerobic tanks of waste treatment plants, as well as in the intestines of certain animals and insects. Through a process known as **methanogenesis,** a group of anaerobic bacteria belonging to the domain Archaea release carbon in the form of methane, much of which finds its way into the Earth's atmosphere. It has been estimated that hundreds of millions of tons of methane are released worldwide each year as a result of microbial activity, 10 million tons of which are contributed by landfills in the U. S. alone. The release of this gas presents a number of environmental problems.

Since it is flammable, methane can pose a serious risk of fire or explosion where it collects. For instance, if special precautions to prevent the gas from accumulating are not taken, proposals to place buildings on abandoned landfills have usually been rejected due to the hazards involved. In most modern landfills a system of pipes is buried among the refuse to collect the methane, which can then be purified and used as fuel or safely vented into the atmosphere. Because most landfills do not have these features, the release of methane by U. S. landfills adds to the atmosphere's greenhouse gases, which are thought to be responsible for global warming. Unlike carbon dioxide, methane released into the atmosphere is essentially unavailable as a carbon source for plants and other organisms except for a select group of bacteria called **methanotrophs** ("methane eaters"). Being aerobic, these organisms are found in the uppermost levels of soil. As methane that is produced in the deeper soil layers diffuses upward, a certain amount of it is intercepted by the methanotrophs and used as an energy and carbon source. While some methane still escapes, at least a portion of the gas is prevented from entering our atmosphere. Actually the carbon in the methane is not entirely lost. Some of the methane that does make it to our upper atmosphere may be converted to CO_2 by reacting with so-called hydroxyl radicals (OH) that are produced through solar powered photochemical reactions. That CO_2, if it makes it back to Earth's surface, could be taken up by plants and other photosynthetic organisms and again become part of the carbon cycle.

The Stuff That Doesn't Go Away

Not all matter that ends on the ground is as rapidly decomposed as our leaf. Glass bottles and ceramic dinnerware are obvious examples of objects that can last for thousands of years. Synthetic polymers commonly known as plastics are another example. While plastics do undergo limited physical and chemical destruction in the environment, in spite of their basically carbon structure, they generally are not readily susceptible to microbial attack. Microorganisms simply do not have the enzymes to break down most plastics. A Styrofoam coffee cup may remain where it was discarded for many hundreds of years, for example.

Researchers have developed biodegradable materials that might replace plastic ones for certain applications. For instance, a biodegradable substitute for Styrofoam, made from limestone, potato starch, recycled paper and natural waxes, might be used for fast food containers, which are usually quickly used and discarded. Under ideal conditions the new containers biodegrade in less than a year. Another significant advantage of biodegradable polymers is that they are mainly composed of renewable resources while ordinary plastics are manufactured from fossil fuels. Biodegradable polymers have been used to make disposable toothbrushes, flatware, bottles and other items. Cost has been the limiting factor in wider applications of biodegradable polymers. Consumers generally have not been willing to accept the increased cost of a biodegradable product even though it may be more environmentally friendly.

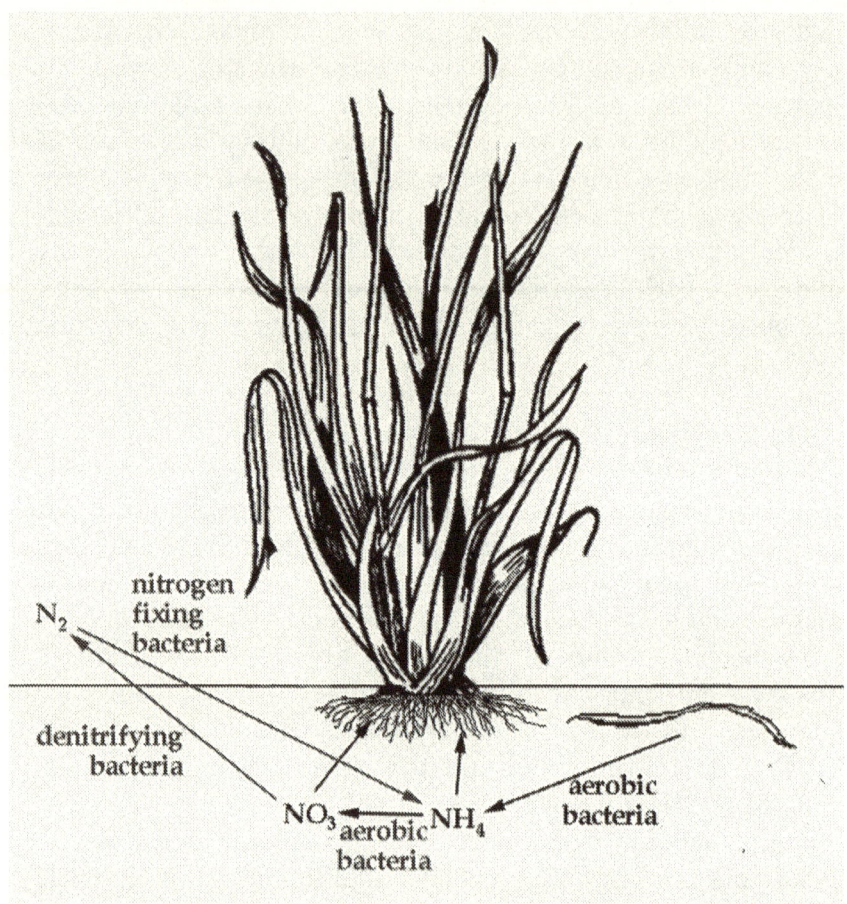

Figure 5.2 The nitrogen cycle.

Nitrogenous constituents (proteins, nucleic acids) in the fallen leaf are broken down by soil microorganisms and converted into ammonium and nitrate, key nutrients that are taken up by plant roots. Under anaerobic conditions, nitrate can be converted into gaseous nitrogen (denitrification). Atmospheric nitrogen (N_2) is converted into ammonium through nitrogen fixing bacteria.

The Nitrogen Cycle.

Nitrogen is one of the most critical chemical constituents that make up all living organisms. Nitrogen is found in all the "working" molecules of a cell, such as proteins (including enzymes) and nucleic acids, as well as those molecules that fill important supporting roles, the vitamins. Life is not possible without a ready supply of nitrogen. Being large polymers like cellulose, the proteins and nucleic acids in our decomposing leaf must be reduced to small subunits before they can be taken up by the microbial cells (Figure 5.2). Enzymes released by the microorganisms, known as proteases and nucleases, accomplish this by breaking down proteins and nucleic acids into amino acids and nucleotides, respectively. These smaller subunits can then be taken up by the microorganisms to be reassembled for their own proteins and nucleic acids.

If the supply of nitrogenous subunits such as amino acids surpasses a cell's needs, the subunits may be further disassembled to salvage their carbon skeletons. In the process, the unwanted nitrogen is removed, usually in the form of ammonium ion (NH_4^+), which is released into the surrounding soil. Here it can experience several fates. Some of it may be oxidized to ammonia (NH_3) or nitrate (NO_3) by other soil microorganisms. Ammonia is volatile and may escape into the atmosphere. Some of the ammonium and nitrate may be utilized by the microbes as a source of nitrogen while much of it can be taken up by nearby plants through their roots to help make more plant structure.

While animals can utilize a variety of organic sources of nitrogen, plants can only use two inorganic forms of nitrogen, ammonia (or ammonium ion, NH_4^+) and nitrate ion(NO_3^-). Both being highly soluble, these nutrients do not last in the soil for very long. They are often rapidly leached by rain or irrigation that carries them deep into the soil and out of reach of plant roots. Thus plants' only source of nitrogen must be replenished continually in order to support their growth. In the absence of fertilization, microorganisms in the soil must perform this function.

The Rhizosphere

The portion of soil that is influenced by a plant's roots is known as the **rhizosphere**. The rhizosphere is where most soil microorganisms, mainly bacteria and fungi, are found. One gram of rhizosphere soil might contain as many as 10^9 CFUs, 100 times more microorganisms than normally found in the surrounding soil outside of the rhizosphere. The microbes are attracted to the rhizosphere by chemicals released by the roots. The kinds and numbers of microorganisms within the rhizosphere have a profound influence on the health of the plant. These microbes support the plant in a variety of ways, including supplying it with plant growth hormones and various nutrients in usable form (such as ammonium and nitrate), protecting it from pathogens through competition, and helping retain moisture. In return, the plant provides the microorganisms with a rich selection of organic nutrients like amino acids, sugars and vitamins, which are released into the soil by the roots. Once again we encounter an example of mutualism operating in nature. To improve soil conditions, farmers can add commercial preparations to their soil that contain live **mycorrhizal fungi**. These products contain mixtures of fungi that support the growth of the farmers' crop plants.

Nitrogen Fixation

Air is about 78 percent nitrogen (N_2), 21 percent oxygen (O_2), and the balance consisting of various gases such as carbon dioxide (CO_2) and argon (Ar). In spite of its importance for life, the N_2 that is in our atmosphere is not directly available to plants and animals. However, early in the course of microbial evolution, some bacteria acquired the ability to convert atmospheric nitrogen to ammonia (NH_3) through a process known as **nitrogen fixation**. As mentioned above, ammonia is a ready source of nitrogen for plants. Over 100 species of bacteria are capable of fixing nitrogen. Most of these organisms are found in the soil, where they can supply plants with that critical element.

Some nitrogen fixing bacteria have developed a unique form of mutualism with certain plants, while others have evolved similar connections with animals such as shipworms and termites. Certain nitrogen-fixing bacteria are able to infect the roots of **legumes**, plants that form seed pods. Examples of legumes are alfalfa, soy beans and peas. While plants normally would defend themselves against invading bacteria, infection by nitrogen fixing bacteria actually is encouraged by the plants through the production of various chemicals that attract the microbes.

Once the nitrogen fixing bacteria have contacted and penetrated the plant roots, they begin to grow within the plant root tissue and produce a tumor-like structure known as a **nodule**. Throughout this remarkable association, two-way communication between the plant and the bacteria is conducted by a constant exchange of chemical messages. The bacteria begin to supply the plant with the ammonia produced from atmospheric nitrogen. In exchange for the ammonia, the plant rewards the bacteria by providing them with shelter, energy-rich carbohydrates and other nutrients. An abundant source of energy is critical, for nitrogen fixation requires enormous amounts of energy to break the bonds that hold the two atoms of nitrogen together. (Remember atmospheric nitrogen (N_2) consists of two atoms of nitrogen whereas ammonia has only one (NH_3).) The bacterial enzyme responsible for nitrogen fixation, **nitrogenase**, is under the influence of a control system similar to those we described in Chapter 1. Should a plant infected with nitrogen fixing bacteria find itself in heavily fertilized soil rich in ammonia, nitrogen fixation is shut down.

It has been estimated that in one year over 400 pounds of nitrogen are fixed per acre of alfalfa. Farmers often grow a crop of legumes and plow it back into the ground just to increase the soil nitrogen content for the next crop, a practice that goes back to pre-Christian Roman times. There are other species of bacteria that can establish symbiotic nitrogen fixing relationships with non-legumes such as red alder, California lilac and sea buckthorn.

Many free-living or non-symbiotic species of nitrogen fixing bacteria are found in fertile soil. All together, microbial nitrogen fixation represents an enormously critical link in the earth's nitrogen cycle and is considered the second most import biological process after photosynthesis.

The Recovery of Mount St. Helens

The spectacular eruption of Washington State's Mount St. Helens volcano in 1980 devastated hundreds of square miles of nearby forest. While deep snowbanks and underground burrows protected small numbers of plants and animals, the blast essentially wiped clean the area's biological slate. Scientists were eager to observe how Nature was going to recover from that cataclysm. They assumed the first signs of new life in the volcano's ash-covered slopes would be microorganisms, lichens, mosses and other simple forms that would be able to establish themselves in the heavy, nutrient poor volcanic residue. To the scientists' amazement, the first significant pioneer organisms to appear were flowering plants known as lupines. Winds apparently dispersed seeds over the barren landscape following the eruption, but what did the seedlings use as a source of nitrogen?

The winds also brought enormous numbers of insects and spiders, most of which failed to survive in the inhospitable ashen environment. Their microbiologically decomposing bodies provided the nutrients that supported the lupines' early development. Once established, the lupines could take care of themselves, for lupines are legumes, capable of symbiotic nitrogen fixation thanks to nitrogen fixing bacteria that had also arrived on the winds. When the lupines completed their growing season, died and underwent their own decomposition, they acted as nurse plants by supplying nutrients for other plants. This progression ultimately resulted in the rapid development of complex plant communities, which in turn supported a wide variety of animals. Also, the millions of trees that were downed by the volcanic blast have been decomposing over the years, adding additional nutrients to the soil. Today the lupines share the mountain's slopes with a scattering of 20 foot western hemlock, Pacific silver fir and countless numbers of other plants and animals. In time, Mount St. Helens will once again be covered by a lush alpine forest, thanks in the beginning to microorganisms.

Denitrification

Not all microbial activities in the soil are necessarily beneficial. Certain species of soil bacteria carry out a reaction called **denitrification** in which nitrate is converted to nitrogen gas (N_2). The nitrogen escapes to the atmosphere and, as noted earlier, becomes unavailable as a direct source of nitrogen for plants and animals. Denitrification is obviously harmful to plants, for it robs the soil of a readily usable supply of nitrogen. Also, the denitrifying bacteria produce an intermediate product, nitrous oxide (N_2O), a gas which also can escape into the atmosphere. Nitrous oxide is a greenhouse gas over 300 times as powerful as CO_2.

The majority of microbial enzymes that catalyze the steps involved in denitrification are sensitive to oxygen, meaning the reaction normally must occur under anaerobic conditions. Several factors can bring about anaerobic conditions in soil: compaction due to excessive foot or vehicle traffic, heavy soil structure (high proportion of clay), water saturation from over-watering, poor drainage, or a combination of these.

Clearly to avoid denitrification in farm or garden soils, air must get into the soil. This can be accomplished by avoiding the factors listed above. For example, soil structure and drainage can be improved by thoroughly mixing large amounts of composted organic material or rock dust deeply into the soil prior to planting and by irrigating plants only when needed. Well aerated soil not only inhibits denitrification but it also encourages root development and stimulates the activities of beneficial aerobic soil organisms. In short, it keeps plants healthy.

Ironically however, denitrification, while directly harmful to plants, actually performs a wider, beneficial role in Earth's nitrogen cycle. Microbial denitrification replaces the N_2 that has been removed from the atmosphere through nitrogen fixation. Scientists believe that if it were not for denitrification that has been going on for over 2 billion years, nearly all of the N_2 that Earth started out with would have been depleted millions of years ago by nitrogen fixation. As we will see later in this chapter, denitrification also has its usefulness in mitigating certain types of soil or water pollution problems.

Metal Cycling

The metallic elements such as iron, aluminum, magnesium, cobalt, etc., are common constituents of Earth's chemical makeup. Some metals such as iron and magnesium fill essential biochemical roles in living cells, while others have no known biological functions. As we learned in Chapter 2, some metals like silver and mercury are toxic to living cells, and a few, such as copper and zinc, are essential nutrients at low concentrations (they help enzymes work) but toxic at higher concentrations. While metals are a normal part of nature, at times human activities have caused toxic concentrations of metals to build up in the environment with serious results. Metal processing, the burning of coal, mining, the manufacturing of batteries, paints and electrical equipment and the dumping of items containing toxic metals such as outdated computers are just a few examples of human activities that contribute to metal pollution.

Microorganisms have evolved various mechanisms to protect themselves from toxic metals in their surroundings. In some instances the microorganisms transform the metal to a less toxic form. Mercuric ion (Hg_{+2}) is the most toxic form of mercury and also the most common form in contaminated soil. It is transformed by bacteria into less toxic, elemental or liquid mercury ($Hg0$) which escapes into the atmosphere by volatilization. Bacteria also have the ability to transform mercury into another volatile form, methyl mercury. While its volatility makes it more easily removed from the environment, methyl mercury is still highly toxic for animals, including humans. Being fat-soluble and therefore subject to biomagnification, methyl mercury often finds its way up the food chain with disastrous results. It is a neurotoxin in humans, causing blindness, loss of speech and hearing, and occasional death. In the last several decades, fish contaminated with methyl mercury have been implicated in a number of mercury poisonings around the Great Lakes, Brazil and most notably in Japan. There, several thousand citizens living near Minamata Bay fell to a highly debilitating disease caused by consuming fish tainted with methyl mercury. A nearby factory had been dumping mercury into the bay for many years. The disease was eventually named Minamata Disease.

Thus it is clear microorganisms in the environment interact with toxic metals primarily to protect themselves. Sometimes these activities put humans and other animals at greater risk, but we also can use microorganisms to help us mitigate metal pollution. This will be covered later in this chapter.

Organic Fertilizers vs. Inorganic Fertilizers

Since plants can use only inorganic forms of nitrogen, phosphorus, sulfur and other elements, it stands to reason the quickest way of feeding plants is to use inorganic fertilizers such as ammonium phosphate or ammonium sulfate. But as noted above, these chemicals are soluble and do not remain in the rhizosphere for long. There are, however, slow release inorganic fertilizers that remain available for extended periods of time. In some of these products the fertilizers are simply less soluble forms of the chemicals mentioned above, and in others the fertilizers are encapsulated in beads to allow slow release of the nutrients.

On the other hand, organic fertilizers and other soil additives such as composted sludge, steer manure, chicken guano, earthworm castings and the like must depend on microorganisms to convert the organic nitrogen, phosphorus and other elements into inorganic forms that plants can use. This takes time and hence the nutrients remain in the soil longer, becoming available to the plants over extended periods of time. However, in reality plants can't tell the difference where their nitrogen, sulfur or phosphorus come from.

Biological Soil Crusts

In many parts of the world, particularly semiarid and arid sectors, undisturbed soil, including sand dunes, will seem to have a crunchy upper layer when walked upon. This is a **biological soil crust,** a cohesive sheath of complex microbial communities in which soil or sand particles are held together by organic secretions formed by certain microorganisms. The crust may be as much as 10 centimeters thick, although most are much thinner. Several microbial species make up a biological soil crust. Major inhabitants are cyanobacteria (photosynthetic bacteria) and algae, as well as fungi, other bacteria, and a variety of mosses and lichens.

Lichens are composite organisms consisting of algae and fungi living symbiotically. In this association the fungi, because of their filamentous nature, act like sponges, capturing moisture from condensed dew. Through photosynthesis the algae in turn supply the nutrients for this odd couple.

Soil crusts provide a shell that protects the soil from erosion by wind and rain, but the crusts are easily damaged by vehicle and heavy foot traffic. The wide use of off road vehicles (ORVs) in isolated desert areas is a major contributor to the destruction of these crusts.

David M. Carlberg, Ph.D.

Composting

Composting is a biological process which allows organic waste to decompose into something that is useful and more compatible with the environment. The organic waste can be sewage biosolids, garden trimmings, agricultural waste, kitchen scraps, or food manufacturing waste. If these materials are deposited in the environment uncomposted, they can cause a variety of problems. Many of these waste materials have a very high BOD content. In addition, untreated biosolids release noxious odors and may contain human pathogens and heavy metals. The major problem with garden and agricultural waste is that it occupies excessive space if deposited in landfills and may contain plant pathogens and damaging insects, while food waste attracts insects and rodents and may also emit noxious odors.

Various food, industrial and agricultural operations produce enormous volumes of waste, much of which is well-suited for composting. Most households also produce good material for composting, such as table scraps and garden trimmings.

Basically in composting, organic waste is placed in piles, windrows (when dealing with large amounts (Figure 5.3)) or bins of some type for smaller quantities. A household compost bin can consist of a simple box-like structure (Figure 5.4). Such bins can be constructed at home from scrap lumber or are available from commercial sources. Three to four feet on a side are optimum dimensions of a compost bin. The most important design feature of a compost bin are openings to allow air to enter from several points, but still be sufficiently enclosed to retain moisture and protect it from rain. Proper aeration for large scale composting is provided by either installing aeration pipes in the heaps, or mechanically turning the material at periodic intervals.

Figure 5.3. A commercial compost operation

a. A commercial composting operation in which organic waste is piled into long windrows. To improve aeration and stimulate decomposition, the apparatus periodically turns over the windrow (b). Additional nutrients and moisture, if needed, are also added at this time. Photo courtesy Midwest Bio Systems/Greg Berry.

David M. Carlberg, Ph.D.

Composting is much more efficient if the starting material is reduced to small fragments. This can be accomplished with the use of a mechanical chipper/shredder(Figure 5.5), or one can use a lawn mower. However, the waste should not be reduced into excessively small fragments, which when wetted causes caking and inhibition of aeration. Fragments from about 1/2 inch to an inch and a half in size are considered optimum. Thoroughly mixing smaller particles such as grass clippings with wood chips, for example, is one way of balancing particle size. In fact the wider the variety of starting materials used the more successful the compost will end up. Table scraps, even cereal boxes and paper egg cartons can be used if shredded, but meats should be avoided. Manure from herbivorous animals can be included but is not essential.

Figure 5.4. Composting bins at a private home.

Various types of organic waste undergo decomposition at different rates, and the end products may differ in quality. One can predict the success of a given batch of compost by a number of criteria, but one of the best is to estimate the **carbon to nitrogen (C/N) ratio** of the starting material. Table 5.1 shows the C/N ratios of a variety of materials commonly used in composting. The ratios range from 4 for fish processing waste to 225 for pine sawdust. The more carbon a material contains relative to nitrogen, the higher the C/N ratio. As a rule of thumb, starting material with C/N ratios between about 25 to 35 have the best balance between carbon and nitrogen and usually experience the fastest and most efficient composting. By mixing materials of various C/N ratios, optimum compositions can be produced. For example, combining equal parts of cattle manure and chopped corn stalks will produce a starting material with an good C/N ratio. Alternatively a small amount of fertilizer such as ammonium phosphate can be added to the compost mixture to compensate for a high C/N ratio.

Figure 5.5. A small chipper/shredder for preparing home garden waste for composting.

David M. Carlberg, Ph.D.

The end product of successful composting is a material that is quite unlike what was originally added to the compost pile. Besides being considerably reduced in volume, it is pleasant smelling, sanitary, relatively free of viable weed seeds and insects and stable (won't turn into something unpleasant). The compost may be sifted through a coarse screen to remove large pieces. If then deposited in the environment it will not cause serious harm, and in fact for the gardener it can be quite a beneficial soil additive, improving soil texture and retaining moisture. In addition it has been shown that a thick layer of compost applied to bare slopes will protect them from erosion.

Table 5.1. Carbon:nitrogen ratios of some common materials for composting.

Fish processing waste	4
Chicken manure	7
Biosolids	14
Alfalfa	15
Cow manure	18
Mixed green weeds	21
Grass clippings	27
Corn stalks	42
Mixed paper mill sludge	61
Rice hulls	85
Oat straw	90
Pine sawdust	225

Source: Sylvia, Fuhrman, Hartel and Zuberer,
Principles and Applications of Soil Microbiology, Prentice Hall,
1998. Used with permission.

Why is the C/N ratio important? If one tries to compost a mixture of materials with a high C/N ratio, say 60, the small amount of nitrogen present relative to the carbon will be quickly used up by microorganisms as they decompose the material and the process will come to a halt prematurely. By the same token, if one mixes the material with a high C/N ratio directly into soil, decomposition of the material will essentially rob the soil of nitrogen. As soil microorganisms utilize the carbon, their only source of nitrogen will be the surrounding soil. Plants will begin to show the effects of nitrogen starvation unless the lost nitrogen is made up by adding fertilizer.

A low C/N ratio is equally undesirable. A large excess of nitrogen will trigger the formation of ammonia, which will be lost through volatilization.

The next ingredient in successful composting is moisture. Generally the starting material should be somewhere between damp and dripping wet. The moisture content of a starting compost mixture can be adjusted by sprinkling it with a garden hose, but excessive water inhibits good aeration and must be avoided. Excessive moisture causes anaerobic conditions, which slows decomposition and leads to denitrification and the formation of hydrogen sulfide and other odorous by-products. Sometimes certain starting materials contain excessive amounts of water and have to be balanced with drier materials.

The process of composting consists of three stages in which a large variety of microorganisms participate. In the initial stage, mesophilic bacteria and fungi begin the breakdown of the starting material, whether it is grass clippings, table scraps, manure, corn husks or a mixture. The microorganisms decompose the waste by releasing enzymes into the surrounding material. The most important enzymes for the early stages of composting plant waste is cellulases and pectinases, which are able to digest the major structural components of the plant material. Other important enzymes include amylase, which digests starch, and proteases, which target proteins.

Nutrients released from the initial breakdown of the plant waste support huge numbers of microbes, as many as a hundred million per gram, which metabolize so vigorously that enough heat is produced to increase the temperature of the compost pile significantly. It is not unusual for a compost pile to reach a temperature of 175°F (80°C) or more. It has been reported that some large compost piles have actually ignited from the heat produced. However, the temperature of a compost pile should not be allowed to exceed about 160°F (70°C), which is optimum for the thermophilic decomposing microorganisms that are involved in the second stage of composting. The temperature can be controlled by turning the pile, that is, rotating inner and outer layers of material. Turning also aerates the pile, stimulating microbial growth.

Besides supporting the growth of the thermophiles, the heat that is generated in a successful compost pile also kills most of the pathogenic microorganisms, weed seeds and insects that may be in the starting material. The thermophilic organisms continue the decomposition of the waste until substantial nutrients are used up, which may take several days. Eventually the pile cools, at which time it should be turned over to aerate the pile. Mesophilic organisms that survived the heat or may have been introduced into the pile during turning become active. Potentially harmful byproducts of the thermophilic phase such as ammonia, acetic acid and ethylene gas are further decomposed by the mesophilic organisms during this so-called curing phase. Once the pile has become stabilized, which normally takes 4 to 6 weeks, the end product can be used in a variety of applications such as a mulch, soil conditioner and low grade (low nitrogen content) fertilizer. But like fine wine, compost improves with age and should be allowed to cure for at least another 2 weeks. Store it in plastic trash bags or used fertilizer bags in a dry location and it will last indefinitely.

Large scale composting does have its downside. During the breakdown of the organic matter in a compost heap, large amounts of volatile organic compounds (VOC) and ammonia, precursors to smog, are released along with carbon dioxide, a greenhouse gas. In the Los Angeles basin, for example, it has been estimated that nearly twelve tons of VOC and ammonia are released each day by the region's composting industry. New air pollution regulations are being put in place that would require composting facilities to install means of capturing at least 80 percent of their emissions.

Living Insecticides

The development of chemical pesticides has saved farmers enormous losses due to insect damage, and has prevented millions of illnesses and deaths among human and animal populations by controlling disease-carrying insects, but at a price. The ideal insecticide would have a narrow range, affecting only a few target species while leaving beneficial species such as lady bugs and bees alone, and it would be harmless to warm blooded animals, including humans. No one chemical pesticide possesses all these characteristics, but many biological ones do. Biological pesticides usually consist of insects' natural microbial pathogens, mainly bacteria and viruses. Leaning on the relatively narrow host range of pathogens, farmers can target specific agricultural pests. The first commercially successful biological pesticide was a common soil bacterium *Bacillus thuringiensis*, which is a pathogen of the highly destructive larval stages of butterflies and moths, commonly known as caterpillars and worms. This bacterium, originally isolated in 1902, was first marketed in the U. S. in 1952 and was aimed at insects like the cotton boll worm and the corn borer. *Bacillus thuringiensis*, in addition to being a spore former, produces a crystalline protein within its cytoplasm which, when ingested by certain pests, is converted into a powerful toxin that kills the insect within hours. The toxin is completely harmless to other insects or animals. The commercial product consists of a powder or liquid suspension of *Bacillus thuringiensis* spores which can be sprayed onto crop or ornamental plants. Dipel® and Thuricide® are the names of two examples of commercial products containing the bacterial spores. Other strains of the bacillus have been isolated that are effective against house flies, mosquitoes and beetles.

The application of *B. thuringiensis* ("Bt") on plants does have some disadvantages. It is more expensive than chemical pesticides and since the spores are viable, they are susceptible to the slow but lethal effects of solar radiation and desiccation once they are applied to plants. In one of the most remarkable successes of genetic engineering, microbiologists have been able to avoid these problems by cloning the Bt toxin gene directly into plants. The result is that cotton and corn plants manufacture the toxin as if it were a normal

179

plant protein. When an insect ingests leaves of the plants, it takes in the toxin and succumbs just as if it had ingested the bacteria themselves. And as pointed out above, since the toxin is harmless to warm blooded animals, there is no danger to any animal, including humans, that consumes any part of the genetically modified plant.

But a problem appears to have risen since the general use of the genetically modified plants. Some scientists have observed that a few insects have developed resistance to the toxin in a manner similar to the acquisition of antibiotic resistance by bacteria. If this phenomenon proves to be wide spread, this pesticide that has been useful for so many decades may rapidly become ineffective. This aspect is now under intense investigation.

Microbial Degradation of Concrete and Iron

The deterioration of materials was one of the three antisocial activities of microorganisms that were noted in Chapter 1. Besides their ability to destroy materials such as paper and wood, microorganisms also have the capability of causing the deterioration of such things as concrete and iron. News accounts of failures of sewer pipes and consequent spilling of raw sewage into the soil or nearby bodies of water have seemed more frequent than ever. There are several causes of sewage spills: human errors, electromechanical equipment failures, and pipe blockage or disintegration. When one uses the term "cast in concrete", one usually thinks of something that is irreversible or permanent. Under the right conditions, concrete is nearly as susceptible to microbial breakdown as a pile of grass clippings; it only takes a bit longer.

Domestic sewage, as it flows through a typical concrete pipe, contains high concentrations of sulfur-containing organic matter, which is rapidly broken down by the many bacteria present (Figure 5.6). Their rapid metabolism depletes the oxygen in the sewage and provides anaerobic conditions that stimulate the activities of a special group of bacteria known as **sulfur reducers**. These bacteria convert the sulfur in the organic matter to hydrogen sulfide, the odorous and poisonous gas that one frequently associates with sewage. The hydrogen sulfide disperses out of the liquid portion of the sewage and dissolves in the biofilm that has formed on the moist, inner walls of the concrete sewer pipe. Here a different group of bacteria, known as **sulfur oxidizers**, utilize the hydrogen sulfide as a source of energy by oxidizing it to sulfuric acid. The sulfuric acid reacts with the lime, the binding agent in the concrete, and literally turns the concrete into dust at a rate that may reach 2 or 3 centimeters a year (Figure 5.7). Within a span of 20 years or less, the pipe wall has thinned to the point of collapse.

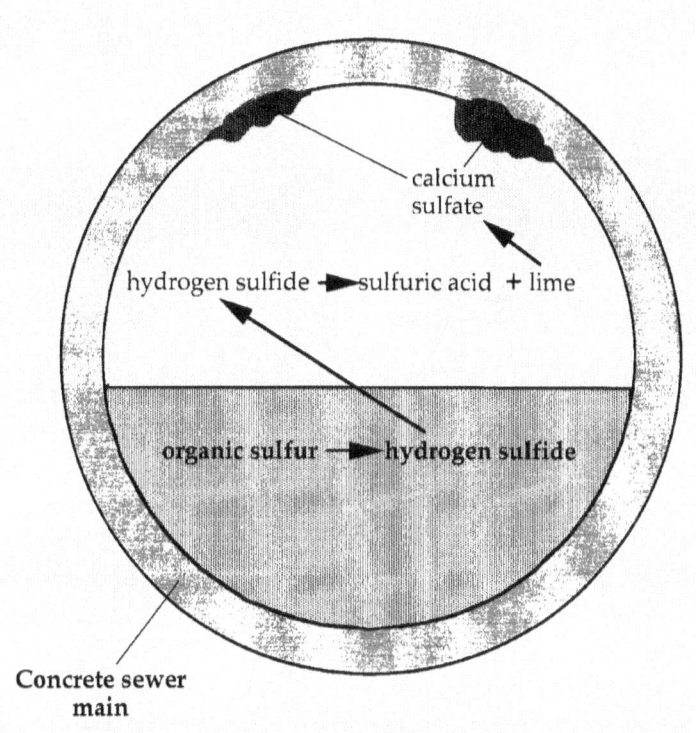

calcium
sulfate

hydrogen sulfide ➤ sulfuric acid + lime

organic sulfur ➤ hydrogen sulfide

Concrete sewer
main

Figure 5.6. Organic sulfur compounds in sewage
are converted by bacteria to gaseous hydrogen sulfide, which rises and dissolves in the moisture film on the walls of a concrete sewer main. Other bacteria convert the hydrogen sulfide to sulfuric acid, which turns the lime in the concrete to powdery calcium sulfate, leading to the eventual failure of the pipe.

One approach to preventing the deterioration of concrete sewer pipes is either reducing the formation of hydrogen sulfide or neutralizing it after it is formed. This also has the additional benefit of reducing odors that often escape into nearby neighborhoods. Neutralizing chemicals such as sodium hydroxide (caustic soda), iron chloride and hydrogen peroxide can be injected at various points along a sewer line. Another approach is lining sewer pipes with a plastic sleeve that shields the inner concrete surface from the hydrogen sulfide.

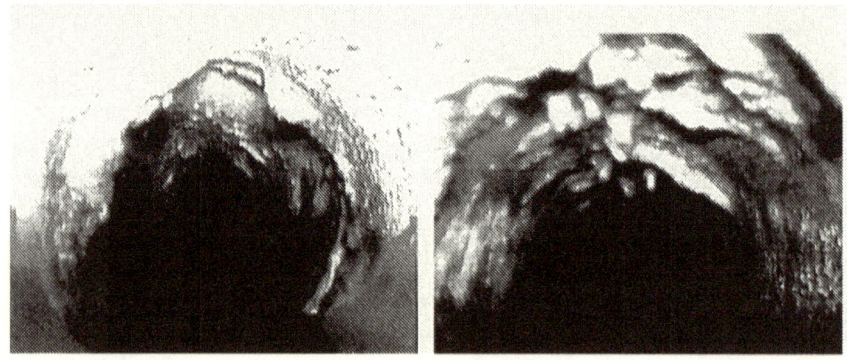

Figure 5.7. Remote camera photos of sewer mains
that are disintegrating as a result of microbial activity. Note the cotton-like accumulations of calcium sulfate forming on the walls of the pipe. Photo: City of Huntington Beach, California.

While it occurs at a much slower pace, when sufficient moisture is present, the deterioration of statues and architectural features made of concrete and related materials is based on similar chemical reactions involving microbial activity.

The same sulfur reducing bacteria also are responsible for the corrosion of iron tanks and water pipes, but a different set of chemical reactions are involved. Here sulfur, which occurs in some waters mainly as sulfate, is reduced by microbial activity, creating an electrochemical reaction that eats away the iron in much the same way the electrode of a storage battery is destroyed as the battery is discharging. Eventually the wall of the tank or pipe thins to the point that it fails.

The plush ocean liner Titanic, which sank in the North Atlantic in 1912 after having side-swiped an iceberg, rests at a depth of about 12,500 feet. Its steel hull is gradually disintegrating not from chemical rusting but from the action of microorganisms. It has been estimated that the 20,000 tons of steel that make up the ship's bow section alone will be completely decomposed in 280 years.

Acid Mine Drainage

During the last 100 years, over 10,000 miles of American rivers and streams, principally in the eastern United States, have been polluted with acid and other toxic chemicals originating from abandoned mines. The practice of strip mining, together with the open-air dumping of millions of tons of mine tailings, had led to this serious ecological problem. Both strip mining and the open deposition of mine tailings expose certain minerals to air and rain, causing the minerals to be converted into powerful acids by chemical and biological processes. The acids, approaching the strengths of battery acid, along with various other toxic chemicals, are then washed by rain into adjacent waterways where they cause enormous ecological damage.

The principal villain in the formation of acid mine drainage is pyrite, or iron sulfide, one of the most common minerals on Earth. Normally protected by a layer of soil, when exposed to oxygen and rainwater, this mineral is converted spontaneously to iron hydroxide and sulfuric acid. This reaction is relatively slow, but as the acid accumulates conditions begin to support the growth of certain acid-loving soil bacteria known as **acidophiles**. The bacteria can utilize the pyrite as a source of energy, producing more sulfuric acid and chemical by-products that actually accelerate the spontaneous chemical oxidation of the pyrite. The combined chemical and biological actions result in the rapid formation of a large amount of acid that leaches iron and other metals from the minerals present. Eventually the acid/metal cocktail flows into nearby rivers and streams where the acid is partially neutralized when it mixes with uncontaminated water. The slight upward shift in pH causes the formation of a variety of insoluble iron compounds (mainly ferric hydroxide) which cloud the waterways with the signature of acid mine drainage, a slimy yellowish-brown sediment (see cover photo). The sediment chokes plants and animals, essentially destroying the water's biological productivity as well as its esthetic appeal. A similar reaction occurs with aluminum, resulting in the formation of insoluble residues of aluminum hydroxide.

One way of preventing acid mine drainage involves sealing the mines or burying the tailings to protect the pyrite from oxygen. Another approach is to add neutralizing chemicals to block the initial accumulation of acid and thereby stopping the acidophilic bacteria from becoming active. A third solution is allowing the mine drainage to pass through wetlands before entering major waterways. Nearly 1000 constructed wetlands (see Chapter 3) have now been established in Appalachia to protect rivers and streams from acid mine drainage. Through a combination of physical and biological action, sediments are trapped, heavy metals are retained, and acids are neutralized in these wetlands.

Mining with Microbes: Microbial Metal Extraction

Ironically, the same bacteria that contribute to highly damaging acid mine drainage can be used to help extract metals from low grade ores. It is not economically feasible, for example, to extract copper from ores containing low concentrations of the metal by conventional chemical methods. But by letting bacteria do the work, the copper can be extracted at low cost. Copper sulfide is one of the most common forms of copper in nature. Acid is added to a giant "leach pile" of low-grade ore to stimulate the activity of certain acidophilic bacteria. The bacteria oxidize the copper sulfide, converting the copper to a soluble form that can then be washed out of the pile and turned into copper metal by allowing it to react chemically with scrap iron. Over 30 percent of all copper and uranium produced in the U. S. is extracted with the help of bacteria.

Oil Field Microbiology

From a distance oil fields seem to be nothing more than a collection of bobbing pumps, pipes and tanks. In actuality much of an oil field's operation may involve battling microorganisms. Many oil field operators pump water into oil-bearing rock to prod the oil out of it, but if the water contains the wrong kinds of bacteria, it could mean trouble. One group of bacteria that oilmen hate are called Sulfate Reducing Bacteria (SRBs), some of the same villains we encountered in the crumbling sewer pipes earlier in this chapter. SRBs role in life is to convert relatively harmless but common sulfate molecules into potent hydrogen sulfide. As you recall, hydrogen sulfide is a poisonous gas that gives rotten eggs and sewer pipes their characteristic odor. Around the oil pumps, the gas is a serious health hazard to oil workers and at low concentrations a nuisance at best to anyone living downwind of the oil field. But in addition, its presence in and around oil wells causes the corrosion of expensive equipment and its eventual failure. Oil field operators must add disinfectants to their injection water to combat the SRBs as well as other microbes that cause the buildup of slime and deposits in pipes and tanks.

Microorganisms can also play beneficial roles in oil field operations. By adding certain species of bacteria to the water that is injected into the oil bearing rock, the oil can actually be removed much more readily. The bacteria produce surfactants, natural detergent-like chemicals that make it easier and inexpensively for the oil to slip through rock much like synthetic detergents allowed the roast pig stains to slip out of the shirt you wore to that Hawaiian luau.

Biodegradation of Toxic Waste

The production of millions of tons of solid and liquid waste has been one of the down sides of the industrial revolution. Much of this waste contains toxic materials which more and more frequently have found their way into drinking and recreational waters and wildlife habitats.

Solid Waste

Ancient trash dumps are one of the richest sources of information of past civilizations and are highly sought after by archeologists. Whenever prehistoric humans abandoned a campsite, they left behind the discarded remnants of their meals, tools, hunting and cooking implements, clothing and other items plus the remains of their dead. The environmental impact of such sites was minimal. In the present day, it has been estimated that over 200 million tons of trash or **municipal solid waste (MSW)**, are discarded in the U. S. each year, or about four pounds per day for every man, woman and child. Such household trash consists of a mixture of paper, yard clippings, metals, glass, plastics, food waste, used diapers and other materials. Over 60 percent (by weight) of MSW consists of biodegradable materials such as paper, yard trimmings and food scraps.

Recognizing a need to control the enormous stream of trash spilling out from our homes and businesses, the EPA has proposed three points of attack: source reduction, recycling, and composting. **Source reduction** simply means redesigning products so they last longer and are not discarded so quickly, and packaging them so less is thrown away. The packaging of food, however, presents a balancing act. As noted in Chapter 2, robust packaging of food products is necessary to protect them from contamination by pathogens or spoilage organisms, and the packaging is usually discarded quickly.

On the average nationwide nearly 30 percent of household solid waste is now recycled or composted, but the remaining is either incinerated or buried in landfills, both of which can cause environmental damage. The open burning of trash is still common in many communities in the U. S. The practice is often carried out in back yard barrel incinerators or open trash dumps. This burning of trash in the open air not only releases carbon dioxide, a greenhouse gas, but also lead, mercury, smoke particulates and toxic combustion products like dioxins. However, most solid waste that is incinerated by commercial operations is done in modern facilities that trap the majority of toxic combustion products before they enter the atmosphere.

Landfills for disposing of solid waste historically have been abandoned quarries, natural canyons, marshes or simply excavations in isolated open spaces. At periodic intervals a layer of soil was spread over the waste to control odors and keep out insects and foraging animals. As the waste slowly decomposed, rainwater could penetrate into the waste and wash soluble material, known as **leachate**, into nearby surface or ground water supplies. In addition, pathogens from disposable diapers, pet feces and food waste could find their way into ground water sources. Finally, decomposing solid waste released carbon dioxide, methane and other greenhouse gases.

Recognizing the potential health and environmental dangers of landfills, the U. S. Congress passed the Resource Conservation and Recovery Act in 1976 in an attempt to tighten standards applying to landfills. The act required several amendments before the legislation became truly effective. The amendment of 1993 set design standards for **Modern Sanitary Landfills.** These are elaborate structures that are constructed to retain the liquid and gaseous by-products of decomposing trash before they enter the environment. A properly designed landfill is lined with a layer of plastic or clay that prevents leachate from penetrating surrounding soil, and once the landfill is up to capacity it is covered with a heavy plastic or clay cap that traps gases and keeps out vermin. Collected leachate is diverted to treatment facilities and captured gases are recovered for fuel. To detect liner failures, monitoring wells are scattered about the landfill, and monitoring must continue for at least 30 years after the landfill is closed. Unfortunately since these design requirements are relatively new, most landfills in the U. S. lack these features.

Ideally much of the MSW deposited in landfills, the food scraps, paper and yard trimmings, should undergo rapid biological decomposition. In reality decomposition of the solid waste in most landfills is enormously slow. In excavating landfills, researchers have retrieved 25 year old food scraps and other organic matter that were well preserved and equally old newspapers that were intact and readable. Because trash in landfills is compacted and layered with soil, anaerobic conditions are quickly established. Lack of oxygen and moisture inhibits most microbial activity and actually leads to the preservation of supposedly biodegradable waste rather than its decomposition.

The enormous Fresh Kills landfill on Staten Island, New York is a clear demonstration of the roles oxygen and moisture play in the decomposition of trash. First used in 1948, the Fresh Kills landfill is sited on a former wetland and is unlined. Through cyclic tidal action, water seeps into the landfill twice a day, bringing with it oxygen to support rapid aerobic decomposition. The end result is much of the biodegradable matter in the landfill has been reduced to a gray slime not unlike partially digested sewage sludge. As an unintended and unfortunate by-product of the landfill, when each tide recedes, an estimated several million gallons of gray leachate are flushed into the nearby Hudson River.

Bioremediation of Environmental Pollution

As many as 80,000 synthetic organic chemicals, several in multi-ton quantities, are produced in the U. S. and other parts of the world each year. Through accidental leakage or spills, deliberate disposal, or in the case of herbicides and pesticides, through normal applications, many of these chemicals end up in the environment. These chemicals, known as **xenobiotics** ("foreign to organisms"), impact the environment in a variety of ways. Many are toxic to plants and animals, some lethally. Other pollutants disfigure the environment by increasing water turbidity or leaving unsightly and unnatural coloration in land forms. Because of their persistence in the environment, these chemicals can travel great distances via rivers and oceans. It is not unusual to detect a particular xenobiotic ten thousand miles from its original point of disposal.

Various solutions to answer the problem of chemical waste deposited in the environment have been used, including incineration or removal of the contaminated soil from the site and moving it to another, less sensitive site. Another approach is bioremediation. **Bioremediation** is the application of microbial processes to reduce environmental contamination. Over 100 major pollution sites in the United States (known as "Superfund" sites) are presently undergoing, or are soon to undergo bioremediation. Bioremediation has a number of advantages over other chemical and physical approaches. It is non-disruptive; the site need not be destroyed in order to save it. It is clean and safe; the method depends on natural processes. And it is less costly compared to other methods. On the minus side, bioremediation is only effective towards certain types of pollutants, and it takes time.

As we have seen in several instances, the enormous breadth of metabolic capabilities of microorganisms gives us a potential tool to eliminate chemical pollution in our water and soil. Under the right conditions microorganisms or their enzymes can bring about the conversion of these pollutants to harmless by-products. Bioremediation can be approached in a number of ways. A contaminated site can be treated directly (known as *in situ* **bioremediation**) as is often carried out with oil spills. Alternatively, when dealing with contaminated groundwater, for example, the water can be pumped through above-ground bioreactors and returned to the aquifer or delivered to consumers. Such a process is called **off site or** *ex situ* **bioremediation. Land farming** is another example of off site bioremediation. Here, contaminated soil is transported to a field, sometimes mixed with microorganisms and nutrients (see below) and plowed under. In time the pollutants are converted to non-hazardous by-products.

In theory, the basic premise of bioremediation is to bring together the chemical pollutants and the appropriate microorganisms or their enzymes in order to detoxify or destroy the pollutants. In many *in situ* bioremediations the organisms are already in the site; it is just a matter of stimulating their metabolic activity. In other cases it may be necessary to apply special organisms or enzymes to the site that are specific for a particular pollutant or group of pollutants. That is known as **augmented bioremediation**. Generally blends (known as **consortia**) of different organisms have proven to be the most effective approach to destroying complex mixtures of pollutants.

In conducting bioremediation, as in preparing laboratory cultures, optimal conditions that support microbial growth and activity must be provided: the proper temperature and pH, an optimum moisture level, the correct concentration of oxygen, or when dealing with anaerobes, the presence of other electron acceptors. The proper nutrients and energy sources also must be present. In the case of the biodegradation of xenobiotics, the pollutants themselves may be the major carbon and energy source for the organisms. Other primary nutrients such as nitrogen and phosphorus, if not available in the site, must be added. The geological conditions at the site also are critical factors that often make the difference between success and failure. These factors mainly involve soil composition, porosity, uniformity and other physical conditions. Unfortunately these factors usually are not controllable at a given contaminated site and may reduce the effectiveness of the bioremediation or prevent it entirely.

Finally, the pollutants themselves must be amenable to biodegradation. They must not be toxic to the biodegrading microorganisms at the concentrations in which the pollutants are found at the contaminated site. Sometimes this problem can be overcome by breeding microorganisms that are resistant to the particular xenobiotics. Under ideal conditions, most xenobiotics are biodegradable, but some xenobiotics that are biodegradable in the laboratory may become non-biodegradable when found in natural sites. Such failures are likely due to less than optimal conditions at the site. And then there are a few classes of pollutants that are not readily biodegradable under any conditions due to their specific molecular structure. These are referred to as being **recalcitrant**. Highly polymerized chemicals or those containing halogens (chlorine, fluorine and/or bromine) such as DDT and PCBs are especially difficult for microorganisms to degrade. This appears to be due to the failure of the microbial enzymes to cleave key chemical bonds that hold these molecules together. The reason for this is either the considerable amount of energy that is required to break the bonds or the inability of the enzymes to reach the bonds due to the shape of the molecule.

Bioavailability is a property of a xenobiotic that is often used to describe the ability of microorganisms to access the chemical pollutant in order to degrade it. The pollutants may, for example, be biodegradable in a laboratory culture but at a particular site become adsorbed to mineral particles in such a manner that they are shielded from microbial attack. Microorganisms must operate in an aqueous environment while many xenobiotics are highly insoluble in water, thus reducing their bioavailability. The addition of non-toxic solvents or surfactants (wetting agents) to a site can sometimes aid in the dispersion of insoluble pollutants and make them more accessible to the microorganisms. As described in the section on oil recovery, some strains of microorganism actually make their own surfactants and have been useful in bioremediation processes as well.

Bioremediation clearly has its advantages, but time is not one of them. *In situ* treatment may take as much as a year or more depending on the type or quantity of the target pollutants. But as pointed above, bioremediation has the advantage of being the least disruptive and least costly method of toxic waste disposal.

Phytoremediation is a newly developed process that involves the use of plants to remove pollutants from soil. When trees and shrubs absorb water through their root systems, they often take up chemicals within the rhizosphere that are dissolved in the water such as metals, pesticides and other types of pollutants, thereby preventing them from getting into underground water supplies. The chemicals are stored in the plant tissue or they could be converted into harmless by-products by the plant's metabolism. Fast growing species such as popular, cottonwood, sunflowers and Indian mustard are most commonly used. Phytoremediation is limited to sites of low to moderate levels of pollution. Studies are underway to assess the effects on insects and other animals that may consume plants that are involved in phytoremediation. At the end of 1999, phytoremediation was being used at about 10 Superfund sites,

Oil Spills

Eleven million gallons of crude oil were released into Prince William Sound, Alaska. when the tanker Exxon Valdez ran aground in 1989. An enormous input of manpower was assigned to the area to remove physically a majority of the oil, but following the initial efforts 350 miles of shoreline were still heavily contaminated with thick, toxic crude oil. Analysis of the remaining oil showed that chemical and biological decomposition of the oil was occurring, but slowly. The biological decomposition of the oil was limited by the sparse nutrient levels in the sound's pristine waters. By applying fertilizers to the contaminated areas, the rate of biological decomposition of the oil as much as tripled.

The Exxon Valdez operation is an example of an **enhanced** or **stimulated *in situ* bioremediation** in which advantage was taken of the presence of indigenous microorganisms in the area, but with additional nutrients supplied to accelerate the microbial activity. Populations of indigenous microorganisms such as those encountered in Prince William Sound frequently consist of complex mixtures of bacteria, fungi and algae that operate synergistically to break down organic pollutants. The decomposition of an organic molecule may occur in stages, with several microbial species each contributing to the process while at the same time exchanging nutrients and even genetic information for the mutual benefit of all the organisms present.

Another example of an application of bioremediation that is under development is in cleaning up fuel oil contamination following floods. Fuel oil lines and storage tanks are often destroyed during heavy flooding, their contents dispersed over large areas occupied by farms, homes and businesses. Once flood waters recede, even minute amounts of fuel oil that have been absorbed by concrete and wooden structures leave foul-smelling residues that are almost impossible to remove by conventional cleaning methods. Often the only solution is to tear out the contaminated structures and replace them. Research is now being carried out to determine the feasibility of using bioremediation to remove the fuel oil residues by applying fuel oil-digesting microorganisms to the structures. Bacteria-containing paper or gel-like material is applied to contaminated surfaces. Preliminary results show that 80-95 percent of certain components of fuel oil is removed within four days to six weeks, depending on the component.

MTBE

Methyl Tertiary-Butyl Ether (MTBE) has been a gasoline additive since the late 1970s when the petroleum industry began phasing out the use of tetra-ethyl lead, a significant source of toxic levels of lead in the environment. MTBE is reported to improve auto engine performance and reduce harmful exhaust emissions by raising the oxygen content of the gasoline. Ethanol, although somewhat more costly, is another common gasoline additive for improving engine performance.

When MTBE began appearing in ground water supplies, the source was traced to leaking gasoline storage tanks buried beneath our neighborhood service stations. Unlike most components of gasoline, which are relatively insoluble in water and don't migrate far in the soil, MTBE is readily soluble in water and moves rapidly through ground water. At best MTBE imparts an unpleasant flavor to drinking water but more significantly, it is a possible carcinogen. Communities throughout the U. S. found themselves closing down one water well after another after detecting MTBE in their water. Leakage from boats and other gasoline-driven recreational vehicles is also a significant source of MTBE in water supplies.

Several methods of removing MTBE from drinking water are being used. Because of its volatility, MTBE can be "stripped" from the water by aerating the water and allowing the MTBE to evaporate. The volatilized MTBE must be captured if required by air pollution standards. Another method involves passing the contaminated water through a bed of charcoal, but this method is limited to water with relatively low levels of MTBE and other organic contaminants.

Bioremediation is a promising alternative method of removing MTBE from ground water. The compound is readily broken down by bacteria and a number of pilot programs have successfully eliminated it from contaminated water.

PCBs

PCB stands for PolyChlorinatedBiphenyl. This term refers to a group of over 200 related chemicals that because of their highly desirable physical and chemical characteristics had been used for decades in an enormous range of applications, including insulation in transformers, fluorescent lamp ballasts, lubricants and hydraulic fluids. Their production in the U. S. had been going on since the 1920s, but was stopped in 1977 when their impact on the environment was fully realized. By that time several million tons had been produced, much of which remains in our communities in electrical equipment that was manufactured before the use of PCBs was banned. Enormous amounts of PCBs also lie in legal as well as illegal disposal sites. Because the natural chemical breakdown of the highly halogenated PCBs is negligible and their biodegradation in the environment is equally slow, it has been estimated that hundreds of thousands of tons of PCBs still repose in the soils and waters of the U. S. The biological effects of PCBs on animals varies; some PCBs are carcinogenic, others disrupt functioning of the liver, endocrine or immune systems, and still others appear to affect reproductive processes. Since they are fat soluble, PCBs are susceptible to biomagnification in fish and shellfish. Because of their widespread distribution, nearly everyone has measurable amounts of PCBs in his or her blood and fatty tissues, but usually at concentrations far below what is considered hazardous.

The elimination of PCBs as well as other recalcitrant xenobiotics in identified sites through bioremediation is an example of a solution that is still in its early stages of development. Some progress has been made in reducing contamination by PCBs at certain environmental sites through bioremediation, but as pointed out earlier, many xenobiotics have proven to be extremely resistant to microbial degradation. Through techniques of genetic engineering work is underway to develop strains of bacteria that have the ability to decompose those chemical pollutants that have so far resisted microbial attack. **Genetic engineering** can be defined as the transfer of the genetic information (that is, DNA) for certain metabolic capabilities of one organism into another. Genes that code for degradative enzymes are being taken from various bacteria and combined into "super bugs" that have enormous abilities to break down a variety of normally recalcitrant xenobiotics. Solutions to many of our pollution problems appear to lie in promising future developments in biotechnology.

Metals

Microorganisms can be used to bioremediate toxic metal contamination in the environment. Many species of bacteria, fungi and algae have the ability to trap toxic metals such as cadmium, copper, lead, mercury and arsenic in their outer capsular layers or biofilm matrix. The accumulation of these metals may reach 50 percent or more of the cells' dry weight. In the bioremediation of water contaminated with arsenic, for example, the water is passed through a tank containing the appropriate microorganisms or their products and as much as 98 percent of the arsenic is removed. Like their ability to convert toxic metals to less harmful forms, microorganisms trap metals in this manner as a form of protection. In order to be used in bioremediation, the microorganisms need not be viable or even intact; dried cell extracts are also effective.

The ability of microorganisms to transform metals such as mercury into forms that are less harmful to themselves also can be applied to bioremediation of polluted sites. In other bioremediation processes, metal pollutants such as arsenic or chromium can be transformed into forms that are more easily removed chemically from the contaminated site.

Unlike bioremediation of xenobiotic contamination in the environment, where organic chemicals are destroyed by microbial activity, the bioremediation of metal pollution only immobilizes the contamination and makes it less of a threat. Thus except in a few cases where the contamination is volatilized, the metals remain in the site bound to the microbial cells. Therefore for full remediation, additional physical or chemical methods may have to be applied to the site to remove the metals. The intervention of microorganisms just makes the process easier.

Nitrates and Perchlorate In Drinking Water

The presence of excess amounts of nitrate in drinking water can be a serious health hazard. While there are chemical methods for removing nitrate from water, it also can be eliminated by treating the water with denitrifying bacteria. As you recall from an earlier section of this chapter, denitrification involves the conversion of nitrate to molecular nitrogen (N_2) under anaerobic conditions. By immobilizing denitrifying bacteria in a continuous flow reactor, it is possible to remove significant amounts of nitrate from the water by passing the water through the reactor. *In situ* bioremediation may also be possible.

While soil usually provides denitrifying bacteria with plenty of nutrients, including energy sources, water in a reactor does not. The problem then is to provide an energy source for the denitrifying bacteria in the reactor without adding more contamination to the water, since it must remain potable. A number of chemicals have been used, including methanol and ethanol, which themselves are toxic but can be easily removed. The best choice appears to be hydrogen gas, which itself is non-toxic (although it is flammable) and the end product is pure water.

Perchlorate is a chemical that is commonly used in missile fuels and pyrotechnics. Since it is highly soluble in water, it frequently has found its way into underground water supplies wherever missiles or their fuels have been manufactured, tested or discarded. Perchlorate is especially hazardous to fetuses and newborn babies, for it impedes brain development. The two most favored methods of disposing of perchlorate in water is by ion exchange and bioremediation. While ion exchange methods merely trap the perchlorate, bioremediation destroys the chemical.

Summing Up

"For much of its history, Earth was a planet of microbes. Even today, microbes dominate the living world in terms of biomass and numbers. Although human activity has altered many features of the Earth directly, significant changes on the land and in the oceans and atmosphere have also occurred indirectly through the impact of humans on microbes. Limiting the scope of future changes will require significant and sustained research on the complex interactions among humans, microbe and the Earth's physical and chemical systems."

From: "Global Environmental Change: Microbial Contributions, Microbial Solutions". American Society for Microbiology. 2001.

Further explorations:

Canyon Country Ecosystem Research Site (CCERS) (Colorado). "An introduction to biological soil crusts." 18 Mar. 2003 <http://www.soilcrust.org/crust101.htm>

Chollar, Susan. "The Poison Eaters." Discover Apr. 1990: 76-78.

Delsemme, Armand H. "An Argument for the Cometary Origin of the Biosphere." Scientific American Sep.-Oct. 2001:432-442.

Dobbins, D. C. "Biodegradation of Pollutants." Ed. W. A. Nierenberg. 3 Vols. Encyclopedia of Environmental Biology New York: Academic Press, 1995.

Rathje, William and Cullen Murphy. Rubbish: The Archaeology of Garbage. Tucson: The University of Arizona Press, 2001.

Sylvia, D. M., J. J. Fuhrmann, P. G. Hartel and D. A. Zuberer. Principles and Applications of Soil Microbiology. Upper Saddle River, NJ: Prentice-Hall, 1998.

United States Environmental Protection Agency. "Municipal Solid Waste Management." 18 Mar. 2003 <http://www.epa.gov/epaoswer/non-hw/muncpl/facts.htm>

United States Environmental Protection Agency. "Composting." 18 Mar. 2003 http://www.epa.gov/epaoswer/non-hw/compost/

David M. Carlberg, Ph.D.

Epilog

"When people are denied access to clean water, soil, and air to meet their basic human needs, we see the rise of poverty, ill-health and a sense of hopelessness. Desperate people can resort to desperate solutions. They may care little about themselves and the people they hurt,"– Dr. Klaus Töpfer, Executive Director of the United Nations Environment Programme (UNEP), September 21, 2001.

The events that occurred just 10 days prior to the publication of Dr. Töpfer's remarks immediately shifted peoples' sense of priorities in an unprecedented way. What had been deemed important was now trivial, and what was trivial was forgotten. Whether people ranked environmental problems as important or trivial prior to those events, the problems quickly slipped off most agendas. Red tides and global warming were no longer relevant, it seemed. But Dr. Töpfer's response to that is clear. Families cannot survive if waters are too polluted to drink or to support fishing, soil too contaminated to grow crops and air too foul to allow children to play outside of their homes. Frustration turns into anger, and anger turns into action that is often violent and directed toward others, is Dr. Töpfer's clear message. Solving environmental problems is more relevant than ever before; environmental problems cannot be ignored.

David M. Carlberg, Ph.D.

Index

About the Author

David Carlberg is professor, emeritus, at the California State University, Long Beach. He received a doctorate in microbiology in 1963, and in the years following has been involved in research covering such areas as biological warfare defense, planetary life detection and biotechnology in both industrial and academic settings. He has been a technical advisor in major Hollywood films and an industrial consultant. Until his retirement he taught university courses in microbiology, bioethics, microbial genetics and physiology and is the author of two books on microbiology. He has been a member of the board of directors and president of an environmental organization responsible for rescuing a 1200 acre Southern California coastal wetland from development and has worked on numerous other environmental causes.